HUMAN EVOLUTION AND ITS UNIQUENESS

C.V. RAO

ATLANTIC
PUBLISHERS & DISTRIBUTORS (P) LTD

Published by
ATLANTIC
PUBLISHERS & DISTRIBUTORS (P) LTD
B-2, Vishal Enclave, Opp. Rajouri Garden,
New Delhi-110027
Phones : 25413460, 25429987, 25466842

Sales Office
7/22, Ansari Road, Darya Ganj,
New Delhi-110002
Phones : 23273880, 23275880, 23280451
Fax : 91-11-23285873
web : www.atlanticbooks.com
e-mail : info@atlanticbooks.com

Chennai Branch
5, Nallathambi Street, Wallajah Road,
Chennai-600002
Phones : +91-44-64611085, 32413319
e-mail : chennai@atlanticbooks.com

Printed in India
at Nice Printing Press, Delhi

PREFACE

The book *Human Evolution and Its Uniqueness* has been painted on a large canvas with broad strokes about the evolution of mankind. The book instead of making an in-depth study of human evolution has rather preferred to generously include the details available in the textbooks of those who have delved deep into the subject. These factual details have been extensively considered and given importance in interpreting man's story from my viewpoint. I entirely take responsibility for such inclusion and express my heartfelt thanks to all those great scientists who have laboured in this field for many years to derive factual contents as presented in their books. I am indebted to all the authors — dead or alive — whom I have referred to.

I convey my sincere thanks to the publisher, editor, composer, and printer as well for their remarkable contributions in bringing out this book.

CHANDALADA V. RAO

CONTENTS

1
INTRODUCTION

Ancient Philosophers described man as a bipedal animal that walks on two legs. Others described him as a talking animal. Humans communicate with each other using language. Still others notably Aristotle termed man as a reasoning animal. Humans reason and react to the situations they confront. Animals react in an instinctive manner. Reasoning is the hallmark of humans alone. However we notice in the nature as well as in laboratory observations that some mammals and primates could exhibit rudiments of reasoning. Humans alone can reason using abstract concepts, that is, conceptual thought is possessed by humans alone.

We want to add one more unique feature to humans. Humans seek values also as guides to their behavior. He has to submit his power of reasoning to the power of his quest for values. Otherwise human society will find itself in a dire predicament and cannot survive. Big fish should eat small fish is what our reasoning tells us. Big fish should protect small fish is what our values tell us. There is an unending conflict between these two human faculties. This is going to be all the more acute in the first half of 21st century, and on a global scale too.

To study this human predicament we have to study his history, prehistory, archeology, cultural and physical anthropology, evolution, genetics, origin of life and about the constituents of living organisms. The atoms of elements of which living organisms were made up were themselves cooked up deep in the bowels of stars. The stars and the galaxies which constitute them were born out of the intensely hot and expanding radiation we now call the “big bang” which took place some 13.7 billion years ago. So to understand man we should have at least a smattering knowledge of these subjects which is no more than what is found in school level text books. So we begin with the big bang some 13.7 billion years ago. After some 3,00,000 years from the big bang the hot radiation cooled down and matter began to form by way of Hydrogen and some 20% Helium. Out of this gas, galaxies and stars began to take shape. All the other elements in the periodic table were not formed along with Hydrogen and Helium. They had their own chequered history. They were cooked up from the primary Hydrogen and Helium in the wombs of the stars, so to say. When a star bursts up, these cooked up heavier elements were spewed into the cosmos to form the raw material for the formation of other stars. Thus the elements in the periodic table were built up over millions of years. Of these elements, all the matter as well as the living organisms were formed on our planet earth. Of these the most stable elements are Carbon, Hydrogen, Nitrogen and Oxygen. These four elements together with some other elements like Sulphur, Calcium, Phosphorus in small quantities form the building blocks of living organisms on our planet earth including humans. So we are literally made out of star dust.

Our Sun and its planets were estimated to have taken shape about 4.5 billion years ago. Life has originated somewhere between this beginning and 4 billion years ago. There were several intelligent guesses as well as experiments to demonstrate how life has originated in this remote past. So far only partial success attended them. Scientists are of the opinion that in the next 20 years it might be possible to produce in the laboratory the most elementary life form that replicates itself, some kind of R.N.A.

Quite early in evolution, blue green algae emerged which was able to produce its food capturing sunlight and taking other ingredients from the surroundings to synthesize carbohydrates with the help of chlorophyll. This process is known as photosynthesis. The emergence of blue green algae changed the scene on the planet earth completely. The by-product of this photosynthesis is Oxygen. Over some millions of years, this Oxygen accumulated in the atmosphere and the atmosphere as we know it today was formed. This Oxygen in the higher altitudes became Ozone and formed a protective shield against the dangerous ultraviolet rays which were lethal to life forms. The ultra violet rays were absorbed by Ozone layer without which the evolution of higher forms of life was not possible. This Oxygen formed an important role in the evolution of life. It is an essential ingredient in the metabolism of all other life forms. So the emergence of blue green algae was an important turning point in the evolution of life on our planet earth.

Life has also divided itself into plant and animal kingdoms quite early in evolution. All those that could manufacture their own food using sunlight, Carbondioxide and water were plants. All those which live on other organisms including plants were animals. Thus the plant kingdom is vital for the survival of the animal kingdom not only as food source but also as producer of Oxygen which is vital for the metabolism of the food the animal consumes. Plant life is the base of the food pyramid. There is a third kingdom of life. They are the fungi which decompose dead organic matter.

The animal kingdom is divided into phylums based on some common features they share. Of these the phylum "Chordata" is of interest to us because we are part of it along with primates, mammals, reptiles, amphibians and fishes. Chordates are also known as vertebrates because almost all chordates, other than the most primitive forms have a vertebral column.

All the chordates share some common features, foremost of them being a backbone composed of vertebrae, connecting a head and a tail. They have bilateral symmetry *i.e.*, a right hand and a left hand side. The water living fishes do not have lungs, instead they have gills to absorb the Oxygen dissolved in water. They also do not have legs, instead they have fins.

In the land vertebrates the head and tail are joined by vertebral column and supported by two pairs of legs attached to thoracic and pelvic girdles. This is the basic skeletal structure which supports the body. The bones are moved and manipulated by muscles. The whole body is covered by a protective skin. The head has a movable lower jaw and teeth in the mouth. There are two eyes, two nostrils, two ears and one mouth in the head. There are one anal and one urinary vents at the rear. The mouth and anal vents are connected by the gut through which food passes. Various juices mix with it on the way to facilitate the digestion of food and the nutrition is absorbed and the waste is excreted. Nutrition is circulated through the body by the circulatory system consisting of blood, blood vessels and heart to pump the blood. Oxygen is extracted from the air by the respiratory system and conveyed to the cells by blood. Waste products of the body are excreted through the urinary system besides skin and lungs and the digestive tract. There is a nervous system to gather information about the surroundings through the sense organs. The information is processed in the brain located in the protective cranium of the skull and suitable orders are passed to the body parts through the nervous system. The glandular or endocrine system secretes hormones which also maintains the body in harmonious balance along with the nervous system. They are responsible for some vital functions of the body like metabolism, growth, sexual cycles etc. A genetical system for bisexual reproduction is located near the pelvic girdle.

Vertebrates in general and land vertebrates in particular share the above life supporting systems. There are some important developments contributed to each stage of evolution that had finally led to humans. They are briefly the movable lower jaw with teeth in it contributed by fishes. This innovation was passed on to other organisms that came after fishes and retained and further evolved by them as it was found to be useful. The fishes have neither lungs nor legs. Some of them developed rudimentary legs and lungs and clambered on to the land to exploit the food sources there. At that time some 250 million years ago the land was mostly covered by mud flats and shallow marshes and giant ferns. There were no mountain ranges and not much temperature variation either. The amphibians successfully invaded the land mass but had to confine themselves to the coastal and shallow areas because they had to contend with the problem of desiccation and they had to return to water to lay eggs. These two problems restricted the amphibians to the shallow marshes and coasts of the land mass. The present day amphibians are the frogs and the salamanders.

The next innovation was the evolution of reptiles from 280 to 63 million years. By this time some mountain ranges began to appear and much of the fern forests have gone under to form the great coal deposits. There was some variation in climatic conditions also. The reptiles evolved much variety to take advantage of these varied ecological niches. The chief developments were however the evolution of scales on the bodies which effectively prevented desiccation. The other development was the amniotic egg.

This has effectively prevented the reptiles from the bondage to water. They were able to roam throughout the land mass of the earth and even some shallow water bodies. They ruled the earth for nearly 200 million years from 280 to 63 million years ago. However they have no temperature control mechanism of their body. Their body temperature varied according to the outside atmospheric temperature. This did not permit refinement of body parts and consequently an active life. They were sluggish and went into a torpor during the cold of the night time. Finally some 65 million years ago they came to an almost sudden extinction perhaps due to some severe change in the temperature of the earth. Present day survivors of reptiles are the crocodiles, tortoises and turtles, lizards and snakes.

Two other groups have evolved from reptiles, which are both warm blooded unlike reptiles. They are birds which have adapted themselves to life in the air. The other group is mammals to which we humans belong. Both birds and mammals developed temperature control mechanisms and keep their body heat at a constant temperature. This enabled them to evolve more sensitive parts to interact with their environment in a better way than reptiles or amphibians could do.

Another important feature of mammals is care of the young after giving birth to them by providing milk. The embryo develops in the uterus of the mother and after maturity is delivered. However it is protected and nourished by the mother by giving milk to the young ones. This is a great improvement from reptiles which lay eggs with the yolk in it providing the developing embryo all the nourishment necessary. After hatching, the young ones have to fend for themselves and many perish in the process. The mammals successfully invaded many ecological niches of the planet earth and have become dominant after the extinction of reptiles.

Of all the mammals the primates are of interest to us. There are some 200 species of monkeys which are grouped together as primates. The primary development they contributed was vision. For the other mammals sense of smell is more acute and is the primary sensory organ signifying an enlarged and elongated snout. In primates the eyes occupy a frontal position in the skull as opposed to the lateral position in other mammals. This allows for the formation of two images in the brain which process visual stimuli which give a three dimensional image with depth perception. As the primates advanced from nocturnal life to day time life the vision has also become colour sensitive. The jumping from one tree branch to another has developed the eye hand co-ordination in the brain which was passed on to humans. This also had developed an opposable thumb to the four fingers. The relative area in the brain has increased. The primates also developed group living, bringing forth only one young at a time and caring and nursing it for a longer time than any other mammals.

All the above traits starting from fishes to apes were also to be found in humans, only in a more refined and developed form since humans are also vertebrates. These differences are however only differences in degree and not in kind. Then what is it that sets man apart from other vertebrates which give their distinctive characteristics. These are Erect posture, being able to walk on two legs and freeing the fore limbs, that is, the hands for exploring his surroundings and finally for making and using tools. Tool use requires conceptual thought which we will discuss later.

Humans could communicate with other members of their species using language and their experience is accumulated over generations. Many other animals could teach their young but only in humans such teaching bridges generations. They could accumulate the knowledge acquired by each generation and add on to it their own knowledge and in turn pass it on to the succeeding generation. The learning is cumulative in humans.

Humans live in social groups depending and co-operating with each other and dividing the labour. The closest to humans, the primates also live in groups but the division of labour and co-operation with each other is very much limited and the number in each group does not exceed a limit. Whereas in humans the degree of co-operation, the division of labour and the number in each social group seem to have no bounds and extending without limit.

Other organisms react to the environment in an instinctive and immediate manner. Humans react in a plastic way giving a delayed response. This delayed response is due to their reasoning capacity based on the store of their accumulated knowledge. This gives them alternatives which necessitate choice, plan and purpose. Where there is choice, there has to be criteria for choice. This implies value judgements leading to moral, ethical and spiritual life. This is the uniqueness of humans which has set them apart from other organisms. The development of conceptual thought is at the root of all these unique features.

What is conceptual thought then. Humans have five important senses on which they depend for their information about outside world. They are taste, touch, smell, hearing and the most important of all vision. Perception is stimulus to one or more of our sense organs. Concept is the memory image of that stimulus in brain. Repeated perception of an external stimulus by any one or a combination of these senses give us an idea or concept of that source of stimulus. A single perception does not produce a concept, repeated perceptions are required for concept formation. A single perception can not form the basis of a concept since perception often deceive us or misrepresent the source of the stimulus. So, for any proper concept to be acceptable it should be repeatedly perceived not only by one person but by a number of normal healthy human beings. Otherwise even hallucinations would pass on as concepts.

We could perceive an object like chair from various angles and distances and under various shades of light. These are like separate photographs. But the concept of chair has a unity about it of which the various separate perceptions help form the concept.

It could be a physical object which is the source of our perception or a quality of that object like its smell or colour or a feature of that object like its shape or size etc. Thus any number of characteristics about an external object could be delineated. When they are isolated and given sound symbols to represent them, we have words for our concepts. If two persons agree on the correspondence of the sound symbols to the concepts they represent, then the two could communicate about the external object without the object being present before them. The same process could be extended to subjective feelings like anger or hunger. This method of communication when extended to a group of people, produced words first for objects, feelings etc. and tense, gender, number etc. later, finally leading to grammatical language.

The root of all these developments is conceptual thought. To be able to visualize objects, feelings, gender, tense, number etc. as belonging to a class and give them sound symbols and communicating through them is the most unique feature of humans. Not only that, if we perform any operations on our mental images of concepts, that would become conceptual thought. In more advanced levels of conceptual thought we use values, ethics, morals, law etc. and use language for conceptual thought. Symbolic logic and mathematics are the highest forms of conceptual thought so far reached by mankind.

This is made possible by the development of a large brain capacity, memory, recall, retention of the images long enough to perform mental operations on them together with the development of the larynx in our throat which produces some 50 phonemes or basic sounds which could be combined to produce a number of words. When these words are joined together into sentences any number of sentences could be produced. Group living starting from a nuclear family and extending to large scale societies organized in nation states is a characteristic of mankind which facilitated this process.

Language is a cultural adaptation and such cultural adaptations are not possible without social living. They have evolved with man. Conceptual thought, language, cumulative tradition, social living evolved concurrently influencing each other.

Man is a product of the evolution of life on this planet Earth. Beginning about 4.0 billion years ago when life is supposed to have originated on this planet Earth, it has passed through several stages of development evolving through those millions of years, each stage contributing some characteristics to human evolution. Without going through all those major stages of development, evolution of man is not possible. What is of interest to us here is his evolution during the last 2 million years when the hominids *i.e.*, semi ape semi man evolved to be a full man *i.e.*, Homo Sapiens.

By about 2.5 million years ago homos became terrestrial, bipedal and also on the way to making crude tools, perhaps of perishable materials in the beginning. Tool making requires a degree of conceptualisation. The early tools for a long time were Oldovan, found near the Olduvai Gorge in East Africa. They were round pebbles chipped at one end to have a cutting edge. There was not much improvement in this technology for a long time. He needed another one million years to become fully bipedal and to develop more refined and varied tools, notably hand axes. This also requires more refined conceptualisation. He had to visualize the end product in his mental image first and also imagine the physical operations required to be performed on the pebble available to him. A lot of trial and error involves in this but nevertheless this is definitely a higher level of conceptualisation than what could be inferred from Oldovan tools.

Erect posture, free hands with opposable thumb to the other four fingers, tool use are the basic characteristics evolved in the earlier stages between 3.5 to 1.5 million years, depending one on the other in a sort of feedback. Body size also increased during this period. The evolution of conceptual thought took place during the last one and a half million years in an accelerated pace as evidenced by improved variety of tool kits. During this stage his brain capacity has increased together with the associational areas between sense organs and locomotor limbs particularly between ear, eye and hand, foot co-ordination areas. His communicating powers through symbolic sounds and ultimately through rudimentary language have developed. They require conceptual thought of a higher order vastly greater than chipping away at some pebbles which is the beginning of conceptualisation.

Now we take a look at the mechanism of evolution and see what factors contributed to human evolution and how they have made it unique giving man certain strengths as well as weaknesses.

Life has properties unique to itself. They reside in its organization, not in its materials but how these materials are arranged relative to each other forming some unique molecular structures. These long molecular strands are known as chromosomes and every species of organism has its own specific number of chromosomes and their own peculiar organization or sequence of the smaller units called genes that make up the chromosomes. Man has 23 pairs of chromosomes and some 30 thousand genes. The number is determined by the genome project scientists. These chromosomes determine the pheno type that is the physical structure and the working mechanism and many other characteristics of the organism.

Every organism has innumerable adaptive characteristics determined by the genes. When a mutation or change occurs in the gene sequence of the chromosome the physical characteristic corresponding to the mutant gene also changes since genes determine the proteins which are the building blocks of the body. When this particular characteristic which was produced by the mutant gene is useful to the organism like the long neck of the Giraffe, then that mutation is a useful mutation. In other words that organism should be able to obtain its food better, protect itself from predators and finally live long enough to produce offspring with the mutant genes. All mutations are not useful. Many useless mutations also take place and organisms with such deleterious mutations die out. Thus the fitness for survival of the species is maintained. This is the process of evolution. The useful mutations are selected by nature itself by testing their usefulness in life. Nature acts as a sort of sieve through which useful mutations pass and useless mutations are discarded.

Organisms evolve to take advantage of an ecological niche, usually a food source. Koala bears of Australia live on the leaves of eucalyptus trees and on nothing else. No other animal survives on eucalyptus leaves. The eucalyptus leaves are the ecological niche of koala bears of Australia. Similarly giant pandas live on bamboo shoots only. Every organism evolves itself to live in a specialised habitat. Evolution is determined by the adaptation the organism could make to the environment or the food source. The key to the adaptive nature of any characteristic of an organism is its usefulness under the conditions in which the organism lives. Adaptation has a known mechanism natural selection, acting on the genetics of populations. Those genes which help the organism survive better and reproduce will quickly spread through the population and those deleterious genes which do not promote the organism's survival will fade out or become recessive.

Natural selection took a different route in humans. When we have noted that all organisms have evolved to take advantage of some natural phenomena, usually a food source, the question what particular phenomena promoted the final steps of human evolution has to be answered. Man took advantage of the causality, the sequence present in nature to evolve during the final stages, particularly when he evolved conceptual thought, word symbols for communication, accumulation of experiences and techniques and their transmission, to succeeding generations using word symbols. All these developments made man unique and they were possible because proto man utilized the causality, the spatio temporal sequence, the repetition, order, rhythm, the pattern present in nature to evolve to full man. Without repetition of the perception, concept formation is not possible. Without causality or spatio temporal sequence present in nature which is repetitive and cyclical, repetition of the perception is not possible. Mankind took advantage of this repetitive character of nature to evolve conceptual thought which made all the difference between man and other animals.

This causality is repetitive, not unidirectional or directionless. Day follows night and night follows day. The moon waxes and wanes every 29.5 days. Seasons follow one after the other and repeat themselves again and again. Seasonal variation is responsible for the abundance or scarcity of food. It is this kind of repetitive and cyclical causality that the evolving hominid took advantage of to evolve its brain capacity to the present level of some 1350 cc from a little more than that of a chimpanzee brain of about 380 cc. Large brain allowed him to have more brain matter and more brain matter allowed him to remember the repetitive causality of nature better and thereby profit from it anticipating the sequence of events.

Thus he was better prepared to hunt the animals that migrate with seasonal changes or use definite routes or water holes etc. He was able to anticipate the behaviour of his prey better than any other predator. Added to this is his weapons, though crude were quite effective. This combination put a potent power in his hands and the other organisms had no chance against him, particularly when he hunted co-operating with others like a pack of wolves.

This dependence on repetitive causality for the final stages of evolution of mankind had given him a uniform genetic endowment, because this repetitive causality exerted similar selection pressure for his adaptation in spite of differences in geography or food source. The selection pressure is on more brain matter and memory power and recall of the memory images and their conceptualisation.

This process is similar all over the planet earth in spite of different situations. Man has used a different method to obtain his food and other wants and this has made him what he is inspite of differences in his food source from one area to another area.

Homo Erectus who evolved by 1.5 million years ago never remained isolated. He was quite a foot loose fellow. He had spread out of the African continent some one million years ago. We have the remains of Java man and Peking man to support this. When populations spread so far afield in other organisms they differentiated themselves into several species which are mutually infertile. However, man remained a single species though differentiated into several races. The racial differences are limited to skin colour, body size head shape and size, nose shape and size, hair and eye colour etc. These are superficial differences in his physiognomy depending upon local atmospheric conditions and geography. They will effect in no way the basic human genetic endowment, that which makes homos humans that is semi ape semi man to full man. This basic human genetic endowment is the same all over the planet earth and in males as well as females of the species homo sapiens that is modern man.

However, racial difference in mankind is a real difference. There is no denying this. Let us understand what is race and what is species. Race exists only among mankind. Species exist in all other life because any slight difference will separate them as species. In mankind the slight superficial differences did not separate them into mutually infertile species. That did not form the basic denomination of separation among them. Human species all over the globe are one and the same partaking in the common genetic endowment of Homo sapiens, even though they have many races depending on the geography and environment of their evolution in history during the last perhaps 500 thousand years or so. Molecular biologists assert that these racial differences took shape during the last 30 thousand years only.

The racial differences in mankind are superficial, limited to physiognomy. They are skin deep only. In animals, birds, insects and other organisms such differences do differentiate them from each other. They do not mate and are mutually infertile. Even if they do mate as in donkeys and horses, the offspring, the mules are infertile. Not so in man. People as separated geographically as the Eskimos in the North and Patagonians in the South could happily mate together and produce healthy human babies. Humans in fact did mate together disregarding racial differences throughout their chequered evolution and remained a single species. The basic human genetic endowment, that which makes homos homo sapiens is the same irrespective of the superficial racial differences. This is because homos took advantage of the causality present in nature to evolve their final sapiens qualities *i.e.*, conceptual thought, making and using tools, communication through language, cumulative learning, co-operative and communal living in large groups, development of culture, ritual, belief systems and value judgements and choices and empathy for each other etc.

These developments require the evolution of the brain, particularly in the memory, recall and image formation and its retention and performing operations on such images and the associational areas of sense organs, particularly sight and hearing. The information received from these organs is associated and transmitted to the locomotor organs for quick reactions as when in hunting. Such are the developments in human brain which were effected during the last two million years and more intensively during the last 500 thousand to 40 thousand years by which time homo sapiens in the form of Cro-magnons have become master hunters and superb artists and spiritualists. Their art is a product of their spirituality. It is sympathetic magic, and propitiation of the spirits for success in the hunt, in all probability. The cave paintings in Spain and France do not suggest any thing more than that.

Other organisms react to the environment in an instinctive and immediate manner. Man reacts in a plastic way giving a delayed response. This delayed response is due to his reasoning capacity based on the store of his accumulated knowledge. This gives him alternatives which necessitate choice, plan and purpose. Where there is choice, there has to be criteria for choice. This implies value judgements leading to moral, ethical aesthetic and spiritual life. This is the uniqueness of humans which has set them apart from other organisms. But there are no clear guidelines to these value judgements or morals, and ethics, aesthetics and spiritual life. They exist in man as potentials to be developed and their development takes place in a social setting. But all societies did not have similar back grounds. As the experience of various societies varied so did the criteria for value judgements, morals and ethics, aesthetics and spirituality also varied. This created friction between social groups as they came into contact with each other. These frictions expressed themselves as hostile and base feelings and passions between different social groups as well as individuals within the same social group.

Knowledge has increased manifold during the 20th century. The first half had seen the developments in physical sciences which have ultimately put the destructive power of thermonuclear weapons in man's hands. The second half has placed the power of missiles and satellites and other space technologies in his control which provided the delivery systems to the thermonuclear weapons. Secondly the developments in molecular biology and genetics gave a clear understanding of the biological processes involved in evolution to such an extent that man could now control his own future evolution as well as that of many other organisms.

How man will use all these powers placed in his hands has to be faced by mankind in the 21st century. What choices and what values will guide him? What kind of moral and ethical systems will influence his choices? What legal and governmental forms could make the decisions and channelise these activities? How mankind is going to use this power that his reasoning capabilities have placed in his hands? He used to control natural forces in earlier years by putting nature to work to benefit him. He may be greedy at times in the short range but in the long range he had to depend on yet another unique feature of his evolution. Subject the power of his reason to the power of his quest for values without which human society will be at odds with itself and cannot survive. This is the problem we have to face in the 21st century even more acutely than ever before and on a global scale.

2
COSMIC EVOLUTION

Man is observing the celestial bodies like the sun, moon, other planets and the stars in the sky long since he had any modern observational equipment like the telescopes or satellites. If we observe the sun rise every day from a fixed place and with reference to an object on the horizon like the peak of a mountain or a tall tree, the sun seems to oscillate over a period of time. The extreme positions taken by the oscillations where the sun turns around are known as summer and winter solstices. They occur around June 21 and December 21 respectively, in the northern hemisphere, and *vice versa* in the southern hemisphere. The summer solstice represents the longest day and the winter solstice represents the shortest day of sun light. The middle position of the sun in its oscillations denotes equal length of day and night, or equinoxes. They occur around March 21 and September 21.

These positions foretell seasons. The seasons mean abundance or scarcity of food. It is all the more significant in the northern hemisphere where there were advances and retreats of the polar ice cap covering much of Europe and North America in ice. It advanced and retreated at least four times during the last 600 thousand years, the last retreat being about 12,000 years ago.

The last stages of human evolution took place during the last 600 thousand years and in the northern hemisphere. The importance attributed to the movements of the sun and its influence on the seasonal changes becomes apparent. It is a matter of survival for the early evolving homo erectus. Homo erectus somehow mastered this and became homo sapiens. This mastering process along with mastery over other regularities of nature helped him evolve a large brain and long memory and all other concomitants for evolving from Homo erectus to Homo-sapiens. No instruments are needed for this except long memory and keen observation. Some ancient archeological structures like the Stonehenge in the South-West corner of England stand testimony to the mastering of these solar movements by ancient people.

During the Neolithic times many agricultural societies mastered the movements of celestial bodies. The knowledge they passed on to us is more or less as follows. The sun starting from summer solstice and going to the winter solstice and coming back to the summer solstice again marks one year. On the earth this is marked by the passage of one cycle of seasons. When counted by patient observation and by trial and error by countless generations, the number of days in a year comes to 365.25 days. The other planet moon with its changes in brightness and darkness also offered a way of measuring the passage of time and seasonal change. But this is more erratic. Patient observation showed that 12 new moons and 11 more days make up one solar year. Since seasonal change depended on the movements of the sun and not on moon, dependence on the movements of the moon to foretell seasonal change proved to be inconvenient. However still some societies follow calendars based on the lunar movements rather than based on solar movements.

Further refinements in the observations of celestial bodies showed that the sky is marked by fixed groups of stars with the entire dome of sky appearing to rotate from east to west from sunset to sun rise, with the stars maintaining their fixed positions in the dome. Further more if we observe the sky at a particular time in the night, say before dawn, we observe a particular group of stars and it appears to move slowly from east to west over a period. Observations were made throughout a solar year and the groups of stars that appeared to move from east to west during the year were divided into 12 groups and they were called constellations and given names. Further there are five more planets which wander across the sky in an east west direction. The sun, the moon and the five planets keep themselves within a limited range and this belt is known as the ecliptic. The positions of the sun, moon, and the planets in the ecliptic are noted relative to each other and the constellations of stars they were found in and a lively field of study, Astrology was developed from it. It is still going strong and many leading personalities of the world are its votaries.

As in many other fields, the Greeks were the systematic observers of Astronomy and no wonder many of the planets, constellations and stars were given names from Greek mythology. Even though other civilizations like Babylonians, Egyptians and others made Astronomical observations, their work did not come down to us. It was mostly lost in antiquity. The Greek philosopher Thales who lived in the city of Miletus around 600 B.C. held the view that the earth was round. Pythagoras and his followers maintained that the earth was spherical and moved through space. Aristotle (384-322 B.C.) whose writings came down to us in more or less good shape gave three proofs for the spherical shape of earth. The eclipses of the moon were due to the earth coming between the sun and the moon and the moon being covered by earth's shadow and it was always round. Secondly the north star appears to be at a higher position from the horizon when observed from northern latitudes than when observed from southern latitudes. Then it appears closer to the horizon. Thirdly when we stand on a beach and observe a ship coming over the horizon, we first sea the sails *i.e.*, the top portion of the ship. As it approaches us we see the lower portions. Aristotle further held that the earth was stationary and the sun, the moon, the planets and the stars moved in circular orbits round the earth.

Alexander (356-323 B.C.) the Greek King and disciple of Aristotle effectively wiped out the barriers between Greece and North West India during his campaigns. He founded the city of Alexandria on the Mediterranean coast of Egypt. After Alexander's death in 323 B.C. his generals parcelled out his empire among themselves and Egypt came to the share of his former body guard Ptolemy. He developed Alexandria and his descendants generously endowed it to make it a great centre of Greek culture and civilization. Trade and learning flourished. It is quite conceivable that many Indian traders, scholars and Buddhist monks visited Alexandria in its nearly 600 years of glorious existence. They surely ought to have carried Greek learning to India, which included Astronomy as well. The 12 Zodiacal signs are identical in Western and Indian systems.

Hipporchus a Greek astronomer lived in Alexandria around 150 B.C. He theorized that the earth was spherical and remained stationary. The sun, moon and the planets circle round the earth in perfect circles. The Greeks were such perfectionists that the heavenly bodies being perfect could go round in perfect circles only. These Greek notions were adopted by Christianity and became part of orthodox belief and became a hindrance to Copernicus, Kepler and Galileo during 16th and 17th centuries. This earth centered system was perfected about 140 A.D. by Claudius Ptolemy who was not related to the Egyptian Kings. During the middle ages until Copernicus overthrew it, his astronomical work the Almagest remained, the standard work. The Arab al was put before the Greek megiste to make it Almagest, the great one.

Greek science and learning died with the burning of the great library in Alexandria in about 415 A.D. by bigots intolerant of pagan learning and culture. The torch of learning, particularly Greek science, Astronomy and Philosophy was to some extent carried on by the Arabs during the middle ages. They learned the numerals and zero and the place values of numbers from the Indians and carried them to the west. They have practically taught the Western world all that they have learned from the Greeks in their contacts with the west in Sicily and Spain. The Christian monks translated the works of ancient Greeks from Arabic to the European languages, particularly Latin. The first flower of the revival of learning in the west was Copernicus. From then onwards the leadership passed on to the hands of the Western world not only in science but in almost all walks of life. Several factors contributed to this transformation. The crusades, the rise of Italian maritime states of Genoa and Venice, the sack of Baghdad which ended the Abbasid caliphate in 1258 by the Mongol war chief Hulagu, all contributed to the decline of Arab power and to the rise of European maritime powers.

Astrology, the prediction of a person's life course from birth to death was quite popular in the Muslim world. There was another need. The Muslims have to turn towards Mecca while performing their prayers. This necessitated the determination of the exact direction in which Mecca was located. This religious requirement needed an exact determination of the latitude and longitude on earth as well as the exact angle of the ecliptic and position of constellations and planetary positions in them. Tables were prepared about these positions with great care and precision. Many observatories were built by Muslim Sultans for this purpose throughout the major Muslim states from Transoxiania to Spain. The Astralobe was invented by them as well as many astronomical concepts like Zenith, Azimuth, Nadir etc. Many stars in the sky also received Arab names. The learning of the Indian numerals along with Zero and decimals along with algebra and trigonometry and analytical geometry developed by the Arabs greatly improved the accuracy of astronomical observations. The Europeans benefited by all this learning. The torch of learning was kept burning by the Muslim world for almost five centuries from 8th to 13th and passed on to the Europeans.

Most of the Arab astronomers were critical of the Ptolemaic theory because it failed to explain many observed facts. Finally the task of dethroning the Earth centered Ptolemaic theory fell to the lot of the Polish monk and astronomer Copernicus (1473-1543). He put the Sun at the centre with the planets rotating round it, but in perfect circular orbits. He was unable to shake off the Greek obsession of perfectionism and circular orbits. That required more accurate observational data on planetary motions.

Such observational data was available with Tycho Brahe who had good observatories in Denmark. Before the advent of telescopes he had the most accurate observational data of the day. But he jealously guarded his observational data and he refused to support Copernicus's theory as the Church opposed it. This opposition of the Church continued during the 16th and 17th centuries also. Tycho's observational data devolved on his assistant Johannes Kepler (1571-1630) only after Tycho's death and he was commended to study the motions of the planet Mars with special attention.

Of all the planets, the planet Mars has peculiar movements, moves back and forth and describes loops. It is difficult to ascribe a circular orbit to such movements. But circular orbit is the most perfect orbit and heavenly bodies being perfect can go only in circular orbits. That was the belief for nearly two thousand years since Aristotle. To propose any other orbit would be heresy. However Kepler stuck to the observational facts and boldly tried an elliptical orbit. It suited perfectly well to the observations of Tycho Brahe. An ellipse has two focii. The sun was at one of the focii. The other planets also follow elliptical orbits but with less ellipticity. Kepler finally found three laws which govern the planetary motions.

While Kepler was formulating his laws, Galileo Galilei (1564-1642) was assembling his telescope in Italy. While Kepler was a theoretician Galileo was a man of experiments. Their work laid the foundation for scientific method. Theory should be based on facts and the predictions of theory should be verified by experiments. This is the basis of scientific method. Both Kepler and Galileo faced the wrath of the Church. Galileo with his telescope discovered four moons circling the planet Jupiter and the phases of the planet Venus. Both helped support the Copernican theory.

In the year of Galileo's death 1642, a great scientist Issac Newton (1642-1727) was born in England. He brought the terrestrial and celestial mechanics under the same laws and principles. He showed that projectiles and planets obey the same Laws. For this purpose he invented integral calculus. He proposed three laws of motion. The first one is on inertia. It states that a body at rest or in uniform motion in a straight line will continue to be so unless disturbed by an external force. According to this the moon going round the earth ought to have flown away from earth in a tangential direction to its near circular orbit round the earth. But it is not flying away. It is drawn into a circular orbit by some force. Newton called this force gravity and he also identified it with the force that draws apples to the ground or projectiles and cannon balls to the ground.

What Newton did was the beginning of Mathematical Physics. He applied mathematical principles of reasoning to observed physical facts. When the mathematical equations so constructed based on physical facts and observations are worked out, they lead to new mathematical equations. These resulting mathematical equations depend purely on mathematical logic. However when the resulting mathematical equations are translated into physical facts, they should predict new physical realities which were not known before. These suggest new experiments and when they are performed and the predicted physical facts observed, we have reason to believe that the universe is rational and not chaotic and that its secrets are amenable to human enquiry. Thus the planets Neptune and Pluto were predicted based on the discrepancies in the motion of Uranus. They were duly discovered using a powerful telescope.

The main objective of this book is to establish that humans evolved a rational brain taking advantage of the rationality present in nature. There is order, rhythm, pattern, spatio temporal sequence, in short repetitive causal connections in nature. Newton's work has established this rationality present in nature. Numerous physical facts subsequently discovered also confirmed the rationality present in nature and they were amenable to mathematical logic which is more precise than the older deductive and inductive logic. This required the development of many branches of mathematics.

More accurate observations and developments in astronomy during 18th and 19th centuries more so the propagation of light through absolute space which Newtonian Laws required began to throw doubts on Newton's laws. The experiments of Michelson and Morley on the speed of light could not be reconciled with absolute space which Newton assumed. Finally Albert Einstein (1879-1955) produced his special Theory of Relativity in 1905 and General Theory of Relativity in 1916. He redefined the nature of space and showed that Newtonian equations are not wrong but special cases. Gravity was due to the curvature of space time continuum in the vicinity of matter. He predicted several other physical phenomena which were later found to be true by experimental physicists.

Cosmology the study of the large scale structure of the universe had its beginnings after 1916. For Einstein space time is curved unlike the Newtonian space which was supposed to be flat. Einstein tried to apply some of his equations to cosmology but they did not correspond to the known cosmos at that time. It was not known at that time that the cosmos was expanding. What was known about the cosmos until 1920 was not much more than the solar system and the planets in it. Further developments had to wait until larger and better telescopes were erected. That happened in the decade of 1920s.

The decade of 1920s was a golden decade for the nascent science of cosmology. Theory and experiment progressed hand in hand supporting one another. In the years 1922 to 1927 two men independently showed that the real cosmos would correspond to the equations of Einstein if we assume that the cosmos is expanding. One of the two men was Alexander Friedman, a Russian mathematician and the other was Georges Le Maitre, a Belgian Priest. In the year 1924 the then new 100 inch telescope was erected on Mt. Wilson in California, U.S.A. Edwin Hubble and his assistant Humason established by 1929 that the cosmos was indeed expanding. Thus theory and experiment laid the foundations for an expanding cosmos with a curved space.

If the universe is expanding and if we could look back, it might have had a beginning. In fact Le Maitre suggested that the universe had started expanding after a huge explosion named 'big bang'. It was latter popularized by the Russian born American scientist George Gamow. Before going into the details of the past let us have a brief look into what is in store for the universe and with it for us in the future.

Einstein's General Theory of Relativity did not specify whether the space is positively curved or negatively curved. That has to be established by observational facts only. If the space is positively curved, the universe would be closed. It would curve back on itself. In a closed universe expansion would end due to the gravitational pull of the matter present in the universe and would be followed by contraction. In a negatively curved space the universe would continue to expand forever. The space may have a positive or negative curvature or none at all. The various possibilities depend on the mean density of matter in the universe. This mean density is not known for sure. However current estimates show that there is a lot more matter in the universe than was known previously. Most of it is dark matter.

This means that the universe may be positively curved, unbounded but finite extending indefinitely far in all directions without being infinite in mass. A positively curved universe responding to gravity has to contract again after many more millions of years. After more millions of years of contraction all the matter in the universe would reach a super dense, super hot state dissolving itself into elemental protons and neutrons or even still more fundamental particles, the leptons and quarks. Then the universe might undergo another big bang and start expanding again. This cycle of expansion and contraction might go on indefinitely like palpitating lungs as per some cosmologists for the future of the universe.

However we are on better ground so far as the past of the universe is concerned. Our knowledge is more accurate, supported by observational facts. We are dwelling at length on the past history of the universe because mankind entertained some very romantic ideas about their origin and the origin of the universe and built around such ideas a mythology, belief systems, religion, morals and ethics and behavior norms. When these origins are found to be not so romantic but hard down to earth facts, we should re-examine our belief system based on the facts at our disposal. That is the reason we are dealing with the evolution of the universe and our place in it.

About 13.7 billion years ago the universe exploded in a big bang from a very condensed state. What is it expanding into? Into itself, for space and time also came into being at that instant. It is more like stretching than expanding. English language and our common sense understanding of the word 'expanding' do not adequately describe this phenomena. What was there in the very beginning? We do not know for sure. By about one by ten billionths of a second after the initial explosion the weak and electro-magnetic forces unified into one force. We can test this in our laboratories by creating the temperatures of that moment in the particle accelerators here on earth. The European lab CERN in Switzerland and Fermilab near Chicago, U.S.A. are doing just that. So we can have some fairly good idea of what happened to the universe since one ten billionth second after the big event. By ten millionths of a second there was more expansion and more cooling and elementary particles were formed from leptons and quarks the most elementary particles of matter so far known.

By about 3 minutes, nuclei formed from the elementary particles present, between 3 minutes and 500,000 years there was sufficient cooling and expansion so that electrons and nuclei bonded together to form atoms. The first atoms formed were those of Hydrogen and Helium. Out of this gas further developments took place. The expansion and cooling is going on even now. Slowly the matter came together under the influence of gravity to form the billions of galaxies now and the stars in them and the planets round some stars. Our own milky way galaxy, our own star the sun and our planet the earth and its satellite the moon are all part of this grand unfolding story of the evolution of the universe.

Now do we have any evidence for this? Yes, the universe is cooling since the big bang 13.7 billion years ago. It is calculated that it has cooled down to about 2.7 above absolute zero on the Kelvin scale or –518°F. This would produce cosmic radiation from all sides mostly in the range of short wavelength radio waves or microwaves. It was found in 1965 by two engineers, Arnold Penzias and Robert Wilson working for Bell Telephone Laboratories in New Jersey, U.S.A.

Another evidence is that approximately 25 percent of the mass of universe is in the form of Helium. So much mass of Helium could not have been formed in stars. It was formed between those 3 minutes and 500,000 years when the electrons and protons were coming together to form atoms of deuterium (one proton, one neutron) various forms of helium (two protons one or two neutron) a little lithium (three protons, four neutrons). The percentages of these elements could be calculated by the known condition of the universe at about 3 minutes period. The predictions agreed with observational facts.

In 1992 NASA, National Aeronautics and Space Administration of U.S.A. sent a satellite, Cosmic Background

Explore, COBE, to detect the origins from which galaxies grew out of the cooling gas of the 'big bang'. The COBE satellite detected ripples — regions of the universe where it was colder and some regions where it was hotter. The difference is not much. By mapping the cold and the hot spots COBE produced a photograph of the universe when it was 500,000 years old. In February 2003 NASA released a map compiled by a satellite called the Wilkinson Microwave Anisotropy probe, which shows the universe at the beginning of time. It confirms the big bang theory and gives the age of the universe at 13.7 billion years. Its resolution is 40 times more than COBE satellite and shows the temperature differences of the baby universe much better. The cold spots corresponded to regions where the gas mixture of Hydrogen and Helium was denser than in hot spots. In the vast areas of denser gas the gravitational attraction began to pull the atoms closer to each other. The density of the gas concentration began to increase and in time became a galaxy.

As the density of the gas cloud increased, knots of even denser gas formed. In such knots the force of gravity draws the atoms closer and the contraction increases. As the contraction increases, the pressure and temperature at the centre increases. This increase in pressure and temperature builds up to such levels it results in nuclear fusion. As the nuclear fusion progresses energy is radiated outward from the core in the form of electromagnetic radiation such as radio waves and infrared, then with light a star was born and some millions of such stars compose a galaxy. Our own Milky Way galaxy consists of some 100 billion stars in it. It takes about 80,000 years to go from one end of it to the other end travelling at the speed of light.

The next galaxy Andromeda is two million years away. The galaxies come in groups. Our own galaxy is one of 17 galaxies that is known as the local group. The largest group is known as Hercules. It consists of some 10,000 galaxies each with billions of stars. Latest estimates show that there are at least 125 billion galaxies in the known universe as established by Hubble telescope pictures. If only we could go there and claim them, each of the 6 billion humans on our land hungry earth could have 20 galaxies each! The universe is so vast and it is still expanding. The whole thing is mind boggling. Yet we are part of it. Every atom in our body was cooked up deep inside the bowels of stars that had exploded earlier. Let us see how.

When stars are first formed they consisted of hydrogen and helium only. Their energy which they radiate came from the fusion of Hydrogen nuclei into Helium nuclei. When the star's supply of Hydrogen runs out, it expands into a large cool star called a red giant. But the core of the red giant shrinks further to produce more heat and pressure sufficient for the Helium nuclei to fuse into carbon and oxygen. When the red giant's supply of Helium runs out, it first begins to burn carbon and then oxygen. This process of nuclear fusion goes on upto the elements silicon and Iron. Then it stops as the fusion of Iron do not produce energy but absorb it.

Elements heavier than Iron are formed by neutron absorption. When the star can not have a fusion reaction going on deep inside its core, it loses its balance and the star explodes. During this period which may not last for more than half an hour, intense temperatures are created in which all chemical elements upto uranium are created in a wild orgy of fusion and then blown into space with the explosion of the star.

Such an exploding star is called a supernova. The Chinese astronomers recorded such an occurrence in A.D 1054. In the 1670s another supernova was observed in our own galaxy. The supernova observed in 1987 was in a neighboring galaxy called Magellan. Thus the gas and dust in the interstellar space is made up of carbon and other heavy elements in chemical combination with hydrogen. Out of such dust and gases new stars are born by the force of gravity which draws them together.

Our Sun is not a first generation star. First generation stars consist only of Hydrogen and Helium. The Sun's spectrum shows lines characteristic of heavier elements also. So it could not have coalesced out of hydrogen and helium alone. It coalesced out of Hydrogen and Helium along with heavier elements also sprinkled into space by exploding stars. It is estimated that the sun had formed some 5 billion years ago and it may go on burning its Hydrogen to Helium for another 5 billion years. Then it swells into a red giant. So, we earthlings could expect to last on this planet earth for quite a long time before the swelling sun swallows us.

The cloud of gas and dust containing hydrogen, helium and grains of heavier elements gave birth not only to the sun but its planets also. It is the grains of heavier elements contained in the gas and dust that are responsible for the rocks, the water, oxygen and life itself. Without them there would not have been any life possible on earth. So we are literally made up of star dust.

3

THE EARTH OUR HOME

Of all the Planets going round the Sun, our Planet Earth is unique in many respects. The distance of the Earth from Sun in the solar system gives it an average temperature which supports life. Planets closer to Sun than Earth are too hot and planets farther from Sun than Earth are too cold to support life. Its rotation round the sun on a tilted axis and in a near circular orbit gives it day and night and seasonal variation. The hot interior of the Earth produces convection currents which constantly shapes and reshapes the surface of the Earth. The rain and snow, erode the jagged surface and create the soils which are necessary for plant growth. Its oceans and weather capture the heat of the sun and keep the earth warm and produce rains and snow which are so essential for maintenance of life. All these factors contributed to the origin of life on earth and its evolution during the past 4.5 billion years of its history.

This evolution has finally led to ourselves, humans. Nature has designed us in such a unique fashion, that we can live nowhere else except on earth. The earth is our home. There is no life on the other planets, though scientists suspect that there could be some bacteria like life on mars and one or two moons of Jupiter. But that is yet to be confirmed. Is there life, or intelligent life on other planets rotating round other Sun like stars in our own Milky Way galaxy or other galaxies? There is a probability. But unless there is intelligent life that could send and receive radio signals out into the universe like ourselves, there is no possibility of detecting them. Efforts are afoot to detect such radio signals coming from such intelligent creatures from outer space but the prospects are quite bleak. On our part we have sent spacecrafts showing the figures of male and female etched on a plaque which also showed the position of Earth and Sun and some important radio stars in space. They were carried by Pioneer 10 and 11 spacecraft in the 1970s past Jupiter. Whether they will ever be captured by intelligent life forms is in doubt. Another attempt was made in 1974 by sending powerful radio pulses from the large Arecibo radio telescope in Puerto Rico. These pulses are sent in the direction of Hercules constellation in the hope that its cluster of 300,000 stars might contain a planet which is advanced enough to understand our message and send a reply to us. But the trouble is that this message takes 24,000 years to reach Hercules cluster and another 24,000 years to receive any message they might be able to send back. So it is as good as impossible to either travel or send and receive messages into space. Meanwhile science fiction writers and movie makers are making a killing weaving webs of fantasy.

Leaving that apart let us have a closer look at the uniqueness of the planet Earth. It is not a static planet. It moves. It rotates on its axis first. It exposes every place on its surface to the sun in each rotation. This period of rotation is what we call a day and we divide it into a period of 24 hours. Each and every organism on the planet Earth including ourselves also has a biological clock working inside us. Our bodily functions are performed accordingly. This biological clock is tuned to earth's clock. Humans require nutrition, work and rest in a 24 hour cycle. So do many animals of the tropics. However, we find exception to this rule in the polar regions. There the animals go to hibernation for long periods during winter.

The Earth goes round the Sun in an orbit of 600 million miles. It takes nearly 365.25 days to complete one orbit. This passage round the Sun is responsible for the succession of seasons. Spring follows winter, summer follows spring, autumn follows summer, winter follows autumn, thus making a full cycle of seasons. We call this full cycle of seasons an year. These regular changes of seasons has enormous significance in the life of all organisms including man. They determine the availability or scarcity of food.

What is responsible for seasons then? It is the tilt of Earth's axis from the perpendicular to the plane of its rotational orbit which is responsible to the seasons. This tilt is 23.5°. While the Earth goes on its journey round the Sun, the North Pole is faced towards the Sun around June 21st and away from the Sun during winter about December 21st. Those are the days of maximum tilt and we call them Summer and winter solstices. The south pole faces towards the sun during winter solstice and away from the sun during summer solstice. It is the angle at which the Sun's rays that strike the Earth that is responsible for the amount of heat that is delivered. Seasonal changes depend on the heat received from the Sun. It is in summer that the Sun's rays make the greatest angle nearer to 90° and in winter the least angle. So in summer the heat delivered by Sun's rays is much more than in winter. This makes a telling effect on plant growth and food supply. Plant life is at the base of the food pyramid. All organisms depend on plant life directly or indirectly for their food supply. So the two most important movements of the Earth are its rotation round its axis with a tilt of 23.5° and its annual sweep of 600 million miles round the Sun. Together they give us day and night and the regular seasonal changes.

We have seen that the stars take shape out of dense knots of gas and dust present in the galaxy. Due to the influence of gravity and the rotation of the galaxy itself these knots of gas and dust gradually coalesce together into huge balls which rotate themselves. Out of one such dense gas and dust mass the Sun and its planets took shape. Initially they were much larger as they were loosely packed. But gravity draws the mass in each rotating knot closer together and ultimately the huge ball of Proto Sun began to shine as the force of the contraction ignited the central core with nuclear reactions. These reactions began to emit streams of ions into the outer space. They evaporated away the gases from the surface of the inner planets. The heavier material sank into their core and their size also reduced much. However the planets did not evolve into stars because their mass is not sufficient to produce the intense heat to ignite a nuclear reaction.

Earth's core was heated up by two processes. First the process of shrinking which consolidated the denser material into the core. Secondly the disintegration of the radioactive materials. When there was some let up in these two processes the surface began to cool down and the water vapour spewed up from the interior began to condense into the Oceans. The water also absorbed chlorine from the volcanic eruptions and sodium, calcium and magnesium from the earth's crust which was leached and washed into the oceans by the rains. Thus the oceans were formed and became salty.

What consists of the interior of the Earth is arrived at by indirect evidence and some intelligent guesswork since we can not drill into the core of the earth. Earth has a very heavy density among all planets, five and a half times that of water. But it's crust is made of granites and basalts whose density is less than three times that of water. So what is inside is much denser.

Earthquakes and tremors provide us with much information to study the interior of the earth. Out of these studies seismologists established the following. There are three distinct and separate zones within the earth. The first is the top layer which is called the crust with an average thickness of 30 to 50 k.m. and is mostly composed of light granites of the continents and heavy basalts under the seas. After this there is another discontinuity at 2900 k.m. depth. The material in between consists of 80 percent of earth's volume and is composed of some kind of rock like material whose properties are not well understood. Inside this is the earth's core. It is generally believed to be of molten iron and with some Nickel and Cobalt.

Why we should be studying the interior of the earth? The surface of the earth is sewed up like the patches of a football. They move relatively to one another due to the convection currents inside the molten core of the earth. These movements shape the surface of the earth and the continents over millions of years.

The size and shape of the continents particularly the mountain ranges and their erosion over the years influence the atmosphere and the various climatic zones and living conditions on the earth. These differing living conditions are now called ecological niches. They favour one kind of organism or the other and the organisms evolve to take advantage of that ecological niche. That is the route of evolution.

How the continents are formed is a puzzling question. One plausible answer is tectonic activity. This says that the crust of the planet consists of about eight large plates and a dozen or so smaller plates. They float on the surface of the hot mantle and move about in all directions extremely slowly ranging in speed from 3 cm to 15 cm in a year. Interesting things happen when these plates collide with each other or scrape against each other or separate. There the volcanoes and earthquakes occur.

What happens when, depends on the density of the material of the plates. Generally plates which carry oceans are of basalt whose density is more whereas those under the continents are of lighter granites. So when two ocean plates collide one plate goes under the edge of another plate producing a deep trench. As the rock of the plate sinks into the mantle it melts due to the nearly 2000°F heat of the mantle. Pressure builds up and it erupts in the form of volcanoes under sea and the molten rock comes up in the form of lava. If this continues for some years, a series of island chains will be formed. The islands of the west coast of Pacific ocean ranging from Japan, Philippines and Indonesia as well as the Gulf of Mexico islands of West Indies were all formed in this fashion. In Iceland we could see this formation of island masses due to the eruption of under sea volcanoes even now.

When a lighter land plate and a heavier Ocean Plate move in opposite directions the lighter land plate rides over the heavier Ocean Plate. It scrapes the bottom of the ocean along with its sediments and the material forms into a series of jagged folded mountains with the sea floor sediments often finding a place on the mountain tops. That is the reason we find sea shells on mountain tops. The Ocean Plate goes under into the mantle and melts. The melted lava comes up in a series of mountains inland to the folded mountains on the coast.

A third possibility is when two land plates collide. Neither sinks under the other. We know what happens when two rams collide. They rise up. The same happens here also when two land plates collide. Their edges rise up only on a gigantic scale. The Alps mountains of Europe were formed when African Plate collided into European Plate some 130 million years ago. The Himalayas were formed when Indian Plate collided with the Asian Plate some 50 million years ago. Thus, the Himalayas are the youngest of the mountain chains formed this way. Some of the tall peaks formed as late as 10 to 4 million years ago. Being youngest they are also the tallest. Erosion has yet to wear them down.

One more possibility is what happens when two crustal plates scrape along each other on their sides. This is currently producing the San Andreas fault in California and is the cause of the major earthquakes in that state, the most recent one being in 1994. All such crustal movemets result in violent earthquakes.

Throughout its long history of some 4.5 billion years the earth's surface did not remain the same. That much is certain, though we do not know much of what it looked like during nearly 4 billion years of its early history. We call this Precambrian period and its history is totally obscure for us. No rocks of this period survive as they were subjected to intense heat and pressure by the subsequent movements of the earth and rocks were the most important source to know the history of earth's surface conditions.

Some 225 million years ago there was only one single continent called Pangaea. All the present day continents were break away continents from that super continent. Some 200 million years ago Pangaea split up into two continents. The northern portion consisted of North America, Europe and Asia and the southern portion consisted of South America, Africa, Antarctica, Australia and India. The northern portion was called Laurasia and the southern portion Gondwanaland.

During the course of time the continents separated from each other. It took nearly 145 million years for the continents to separate from each other and another 55 million years to reach their present positions. Although North and South America lie on separate plates they both drifted north west and were joined four million years ago. The Eurasian Plate went north east pursued by Africa ramming it on and off and raising the Alps mountains. The last to separate were Antarctica and Australia. The Indian Plate separated from Africa and Antarctica and Australia some 120 million years ago and moved towards Eurasian Plate at the rate of 9 meters per 100 years to ram into the Asian Plate some 50 million years ago. This raised the Himalayan mountains most of which were formed during the last 10 million years. The Indian Plate is still moving at the rate of 4.5 meters a century making the sub Himalayan Zone a potentially earthquake zone.

As the separation of the continents created new mountain ranges and new seas and thereby new climate zones and new ecological niches, they also provided opportunities for the emergence of new species of organisms and new life proliferated. Each continent became unique with its own unique denizens.

We have seen that mountain ranges are formed when two plates of equal buoyancy collide with each other or one plate goes under the other. In the second case the subducted plate breaks down into pieces and melts and builds up tremendous pressure and heat inside the bowels of the earth. This erupts in the form of lava and volcanoes. When two plates scrape laterally against each other and when the stretching of the rock surpasses a limit then the rock snaps and results in earthquakes releasing tremendous quantities of energy. The earthquakes that occurred in Latur, Garhwal and Kutch in recent years in India are of this type. Devastation and death follow in their wake. There is no way that science could predict them.

These natural calamities churn the interior of the bowels of the earth and bring up what is inside and take in what is on the surface in a cyclical manner. The cycle of course lasts millions of years. This process is known as diastrophism. If that alone is acting on the shaping up of earth's surface then the surface of the earth would have been a much uneven and jagged surface unlivable to many species of organisms.

The other force which opposes this is erosion. If erosion alone were to act on the surface of the earth, in about 25 million years of time all the continents would be reduced to mud flats and even the sea would cover them up. The chief eroding agents are water, wind and ice. The diastrophism and erosion together acting on the surface of the earth has created the biosphere of our earth and the various ecological niches which harbour various species of organisms.

The atmosphere sucks up tons of moisture every year in the form of water vapour. This falls back again in the form of rain, snow, sleet and dew mostly into the seas and some on to the land. The rainfall on land mostly goes underground and some of it evaporates but an estimated fifty thousand billion litres runs off into the seas. This is the flood that is the chief agent responsible for the erosion of the earth's surface and shaping it. Water flows down a gradient. The law of gravity applies everywhere. Mountain slopes have more gradient than plains. So the mountain streams are swift and consequently erode more and carry the boulders, pebbles and sand down the stream on to the plains. It is not only the water that is the eroding agent but the chemicals and the fine sand and grains that carries the gravels downstream.

As the rivers age, streams which were once swift and straight broaden out and turn into old rivers meandering sluggishly across broad valleys. These valleys are generally smooth and flat because of mud and silt laid out by the river. During seasons of high water a river may overflow its banks and inundate the plain, adding a fine layer of silt to the flood plain. It is in such flood plains ancient civilizations took root and flourished.

For mankind erosion is a blessing. It provides the chief ingredient for soil without which the continents would be lifeless. Soil mostly consists of mineral grains. The mineral grains together with the decayed organic matter is an ideal medium for plant growth. Depending upon the nature of the original sediments, the rainfall, the mineral content and the organic matter that builds up, these earth deposits may range from barren dusts to the rich black soils which can support crops in great abundance. Erosion also destroys what it builds. The soils of the plains and valleys are slowly carried down towards the sea and piled up at the river mouths or simply sucked away into the sea by the tides.

Apart from running water, glaciers acted as great shapers of the earth's surface. Erosion by moving ice is limited to a few glaciers now, mostly on high mountains. But during the ice ages which began nearly a million years ago, the polar ice advanced and retreated at least half a dozen times over North America and Europe. These thick ice sheets scraped the earth to the bed rock and acted as huge bulldozers. Much of central Canada is barren and its top soil was deposited in the corn belt of U.S.A. So also the New England states of U.S.A were scraped up and the top soil was deposited off the coast of Connecticut state to build up the Long Island. The retreating ice formed the Great lakes between U.S.A and Canada and many smaller lakes in Canada and New England. Similarly in Europe the Scandinavian countries were scraped up and much of the soil was deposited in Ukraine to form its rich wheat fields.

So the Earth's top soils were made by the eroding forces of water and ice sheets and such top soils are essential for plant growth. Plants are at the base of the food pyramid. All other organisms depend on plants for food and some depend on predation of other organisms. Whatever the case may be plant life is at the base of food chain. Without the eroding forces of nature, plant life would have been impossible on Earth.

Apart from the land masses of the Earth there are two other important zones which are essential for life. They are the oceans and the atmosphere.

The oceans occupy three-fourths of the surface of the planet Earth. Oceans have two major life zones, the coastal zone and the open sea. The coastal zone is the warm, nutrient rich shallow water extending between the shore and covering the continental shelf. This supports much life. Next to this is the open sea. Any light worthwhile penetrates upto 200 meters deep into the sea water. This is where photosynthesis takes place. More than 99 percent of all plant life of sea consists of microscopic plants called diatoms floating in the upper 30 meters of the ocean. There they get sunlight and carbon dioxide dissolved in sea water and nutrients like nitrates and phosphates stirred up by the winds from the bottom of the sea. These diatoms or phytoplankton are the primary producers of carbohydrates, sugars and proteins. They are consumed by zooplankton which are not primary producers but primary consumers. Feeding on them are a variety of consumers like herrings, sardines, anchovies and other small fish. The massive baleen whales also consume them in great quantities.

Ocean waters are also great reservoirs of heat. The heat they absorb from sun's rays unevenly heat up the oceans. This uneven heating and earth's rotation, the tilt of its axis and the winds which also depend on the above three parameters, cause warm and cold water currents as well as warm and cold weather currents. These ocean and air currents together redistribute the heat received from the sun. They influence the climate and the vegetation.

The Gulf current which starts from the gulf of Mexico with warm water passes along the east coast of North American continent and along the North Atlantic and towards the Western European countries. It is keeping those continents warm during the winter. Otherwise those continents would be unlivable with sub Arctic climates. The Humboldt current which starts from the Antarctic with cold water and passes along the west coast of South America influences the rains in North America, Australia, Asia and South Africa. Any imbalance in this current will create havoc in the above continents and their economies. The warm waters of the Gulf current are going into the Arctic Ocean and melting ice. This is having adverse effect on the weather in Canada, North America and Western and Mediterranean Europe.

The phenomena known as El-Nino or Child Christ since it happens around December 25th in the equatorial Pacific ocean near the coast of Peru in South America is such an imbalance. The warm waters of the Western Pacific suppresses the cold current of the Humboldt from reaching to the surface, consequently the diatoms die and with them, the zooplanktons, and the anchovies and the cormorants and the sea birds which depend on them also die in their millions. Once in a while this happens and if it happens too frequently and lasts too long the westward pacific winds weaken and rains will fall at places where they are not expected and at unexpected times and in great abundance in short duration. This causes severe floods. At other places where rains are expected they don't precipitate and cause droughts and hardships to the people. The severe drought we went through in India during the 2002-2003 season is a result of such El-Nino effect. Other parts of Australia, South Africa and Central America also experienced severe drought during 2002. Parts of Korea, China, Southern Russia and Eastern Germany experienced heavy rains and floods and damage.

Oceans exist because the surface temperature of the Earth is between 0°C and 100°C. It freezes into ice at 0°C which happens at the polar region. Above 100°C it evaporates which happens at the equatorial zone. This great evaporation of the sea water is the main source of clouds and weather and rains and snow or the hydrological cycle that sustains life on Earth. Water is a great solvent. All organisms regularly conduct chemical reactions inside their body as a part of their metabolic process. For this water is essential. 60 to 70 percent of all organisms' body mass consist of water. So water is the mother of life and without water there would be no life and ultimately the source of all water is the oceans. The oceans provide us fresh water by the process of evaporation and its precipitation in the form of rain, snow, dew or sleet. This distribution is effected by the weather currents caused by the rotation of the Earth with a tilt on its axis.

The weather, apart from providing us the earthlings with fresh water provides another important service. That is the green house effect. The water vapour, carbondioxide, methane, nitrous oxide and ozone present in the atmosphere are collectively known as the green house gases because they trap the heat reflected from the earth and do not let it escape into the outer space. This trapped heat is what keeps the earth warm at an average temperature of 15°C. Without the green house gases the earth's temperature would be -18°C and the earth would be unlivable. So the green house gases are quite essential for the maintenance of life on earth.

Suppose we increase these green house gases in the atmosphere and their heat retaining capacity is increased and consequently the mean temperature of the earth is increased. What will happen then? The phenomena like the

El-Nino will occur more frequently and with greater severity causing floods, droughts and cyclones that force migration of plant, animal, bird and insect species. Some species which can not adjust to these swift changes will have to perish. All these will have adverse impact on human societies and their economies and agriculture.

Man with his ignorance, foolishness, greed or indifference is just doing that, increasing the green house gases, particularly carbondioxide in the air. Our ever increasing hunger for electrical power is driving us to burn more and more fossil fuels like coal, oil and natural gas to generate electricity. Besides the automobile exhaust fumes are contributing nearly 30 per cent to the CO_2 content of the atmosphere along with other harmful ingredients.

The green house gases are keeping the Earth's temperature variation to within 0.5 to 1.0°C over the past 100 to 200 years. We know from other sources that the Earth's temperature, weather and seasonal patterns and soils and ecosystems did not change much since the last 12,000 years when the last of the glaciations retreated. However any change in the content of the green house gases would alter this balance by increasing the temperature of the Earth beyond the normal limits.

Green plants are the primary producers of photosynthesis on earth and convert CO_2 in the atmosphere with water and other nutrients from the soil to carbohydrates. This reduces CO_2 content in the atmosphere. But man is cutting down forests to appease his ever increasing hunger for wood, pulp, grass land and agricultural land. This is reducing the capacity of plants to absorb CO_2 from atmosphere.

A third source of atmospheric heating is our use of chlorofluorocarbons or C.F.Cs. in refrigerator and air conditioner coolants, spray cans etc., These chemicals when escape into the atmosphere gradually work their way up into the upper atmosphere and there react with the ozone layer and convert it into other chemicals. This is creating ozone depletion and holes in Ozone layer. This Ozone layer is the one which is filtering away sun's harmful energy of ultraviolet rays from reaching the earth. They are lethal to life on Earth, cause serious health problems to humans etc. There are already one large hole in Ozone shield over Antarctic and a smaller one over the Arctic. This ultraviolet energy reaching the earth is also contributing to heating up of the earth's atmosphere.

Thus, human activity is increasing the CO_2 in the atmospehre and depleting ozone in the upper layers of atmosphre. These two are contributing to the heating up of the earth. The cutting up of forests is reducing the capacity of green plants to absorb CO_2 in the atmosphere.

Even if we stop the emission of carbondioxide now it is a long lived green house gas and it could stay up in the atmosphere for another 200 years. Even if we stop using any C.F.Cs now they will be stabilised in the atmosphere in the next 100 years only. As a result some of the changes in the climate systems that might be induced like the melting of ice sheets and rise in sea levels and changes in the ocean circulation patterns would be almost irreversible. If the Gulf stream for any reason stop flowing towards north Atlantic, the ice sheet would again descend down on northern Europe and North America.

What will happen then? Since over the last more than 100 years when we have been measuring and keeping records of global temperature, the decade of 1990s were the hottest and the year 1998 is the hottest followed by the year 2001. The warming is linked to the growth of green house gases in the atmosphere and their increase over the normal levels. This global warming resulted in the rise of global mean sea level, decrease in the sea ice thickness and retreat of mountain glaciers. The frequency of extreme weather conditions in recent years such as floods, cyclones and droughts have shown an increase.

The heat waves of 2001 resulted in the El-Nino conditions which caused unprecedented floods of 2002 in China, Korea, Southern Russia and Eastern Germany. Parts of Asia including several states in India, Africa and United States were affected by severe droughts. The strongest El-Nino conditions of last century occured in 1997-98 which affected nearly 100 million people and caused damage worth 100 billion dollars world wide. Forest fires in Indonesia, Russia and U.S.A are a result of severe droughts. An estimated 20,000 people died in Europe in the heat wave of August 2003.

If these disturbances in weather occur too frequently and with increasing severity, then large scale migration of people and other species have to take place. Species which cannot adjust to such migrations have to die. The rise in Earth's temperature and consequent rise in sea level could shift the areas where people could grow food, some areas might altogether become uninhabitable due to inundation of low lying areas, salination of drinking water sources or change of ecosystem. Some of the low lying port towns will submerge. Much of the prawn and aquaculture areas of the world as well as mangrove forests will also submerge.

The nations of the world met in Kyoto in Japan in 1997 to arrive at a protocol to reduce the green house gases gradually. After much wrangling they did arrive at an agreement. It came to be known as the Kyoto protocol. With Russia signing the Protocol in November 2004 it has come into force in February 2005. USA and Australia have opted out of the protocol. China and India are not obliged to sign it until 2012 since their per capita emissions are very low.

The poor nations of the world were exhorted by the rich nations to limit their population and their aspirations for a better life, since a better life means more cosumption of energy and more consumption of energy means more emission of CO_2 from the power plants. Instead they are advised to grow more trees to protect themselves from the CO_2 emissions from the power plants and automobile exhaust fumes of the rich nations.

The people of the industrially developed nations constitute 22 per cent of the world's population. They use up 85 per cent of world's resources and energy and generate an equal amount of the world's waste products and pollutants and green house gases which are released into the atmosphere.

The green house effect has assumed such importance in the second half of 20th century because the variation in Earth's temperature took centuries during the past 12,000 years. Now it is taking only decades. An increase by 2°C would be catastrophic to the people as well as other living forms. The effect will be experienced during our own, our children's and grand children's life time in the next 50 years. Unfortunately the world's governments are in deep slumber over this major issue of 21st Century. They are not even implementing Kyoto protocol to which they were signatories.

4

ORIGIN OF LIFE

'All living things were created by God' is a belief entertained by many societies. It is intertwined with the religious beliefs of some people. Apart from the religious belief there is another stream of belief which was entertained without any proper investigation. This was the belief in the spontaneous origination of life from decaying organic matter. This was proved to be untrue by an Italian Physician Francesco Redi (1626-1698). He showed that putrefying meat if properly covered would not develop maggots in it. The insects were prevented by the cover from laying eggs in it. So spontaneous origination of life from decaying organic matter was disproved.

Louis Pasteur was a French biologist who lived in the 19th Century. He conducted many experiments with bacteria and concluded in 1860s that bacteria could arise from a pre existing bacteria. It can not spontaneously arise out of nothing. Thus the smallest living thing known then, the bacteria, needs a pre existing bacteria to come into existence by cell division in a suitable medium. It is as complex as the single cells of any other higher organism. Thus was established the genetic continuity of all living matter.

Some argued in the 19th century that life had no origin but was eternal like matter. We now know that this view is not correct. Life is not one of the fundamental categories of the Universe like matter, energy, space and time. They had their beginning at the Big Bang as we have seen. Life as we know today is a manifestation of certain molecular combinations. These combinations could not have existed from the beginning of the Universe, since the elements of which living matter is composed came into being much later than Big Bang. The chief elements of life are Oxygen, Carbon, Hydrogen and Nitrogen. We have seen earlier that elements like Carbon, Oxygen and Nitrogen etc., heavier than Hydrogen and Helium took time to fuse in the core of stars. These heavier elements were not there in the beginning when matter began to coalesce after the Big Bang. It was chiefly Hydrogen and some 25 per cent Helium. The other elements were formed much later. Hence life which was formed by these heavier elements was not eternal but had a beginning in time, which was much later than the Big Bang. So when and where and how are the questions to be answered. So far scientists are not yet able to establish when and how life came into being on our Planet Earth. All that they were able to do was push the event as far back as they could in time.

We have to find out the Physical and Chemical conditions under which life has originated. One sure way is to hunt for the fossils as far back in the history of the Earth as we could go. There are fossils aplenty in the Cambrian rocks which period in Earth's geology started some 550 million years ago. These fossils are of well developed organisms and many with hard shells. We know from our present experience that there are smaller organisms and with softer bodies that may not have left any fossils. Similar organisms may have lived in the Pre-Cambrian Period. It is a long period in Earth's History from its origin some 4.5 billion years ago to the start of the Cambrian around 550 million years ago. This is nearly 3950 million years and nearly 88% of Earth's history. It is rather a dark period so far as Earth Sciences and Life Sciences are concerned because of the meagre evidence available by way of fossils and rocks of this period. The surface of the Earth was also much convoluted, crushed and altered by the upheavals of the earth in later periods. So much so rock formations from Pre-Cambrian period are extremely rare. Nevertheless the search for them continued.

Some of these Pre-Cambrian rocks where the fossils were not destroyed by heat revealed much information. The gunflint chert aged 1.9 billion years and found in Southern Ontario in Canada contained well preserved fossils of microorganisms which are said to be bacteria and algae. The Bulawayan lime stone from Southern Rhodesia (now Zimbabwe) and aged 2.7 billion years is said to be a fossil algal colony. The Fig tree chert 3.1 billion years old and found in Transvaal, South Africa contains micro fossils of bacteria and algae. A tiny ball of carbon which lost all anatomical features found in a rock aged 3.86 billion years may be the oldest evidence of life on Earth. It was found on an island off Greenland.* Life on Earth must have formed still earlier than this date and still remains a mystery and we could say that life appeared on Earth some 4 billion years ago.

To unravel the mystery surrounding the origin of life on Earth , we have first taken the route of fossils and gone as far as 3.86 billion years backwards with the fossil find on the island off Greenland. However this is not anywhere near the origin of life. We are still far away from it. Life has originated much beyond 3.86 billion years ago. We will now take the route of biochemistry, molecular biology and genetics and see if we could shed more light on the secrets of life.

Taking into consideration the genetic properties of living organisms, life could be defined as a group of molecules which are self-replicating and mutating and undergoing Darwinian evolution. Living systems are systems that reproduce, that mutate and that reproduce their mutations. Mutation means slight alteration in their genetic material and these alterations are inheritable. These mutant offspring has the capacity to adapt themselves to the vicissitudes of existence *i.e.,* they are capable of undergoing evolution by the process known as natural selection.

Living bodies largely consist of compounds made of the elements Oxygen, Carbon, Hydrogen, and Nitrogen together with a number of other common elements. They make up proteins and nucleic acids which are the chief ingredients of living things. They have a definite shape and size characteristic of that species. They have a definite organization of their body parts to sustain it.

Proteins are often huge molecules. They are made up of chemically linked building blocks known as Amino acids.

* Richard Monastersky, "The Rise of Life on Earth", *National Geographic*, March, 1998.

There are 20 Amino acids. These Amino acids combine with one another in a variety of ways. The number and sequence in which they combine determine the nature of the protein formed. Proteins are the building blocks of living organisms. They are also the facilitators of specific reactions acting as enzymes. So the specific nature of proteins controls the specific nature of living organisms. The specific nature of proteins depends on the order, sequence and number of the Amino acid chains which make up that protein. So there has to be a mechanism to control this combination and order. These are nucleic acids.

Nucleic acids are compounds of still smaller units called nucleotides. A nucleotide consists of a nitrogen containing purine or pyrimidine base, a sugar and a phosphoric acid. Depending on the kind of sugar involved, two types of nucleic acids can be distinguished. One is Ribonucleic acid (RNA for short) and the other is Deoxyribonucleic acid (DNA). The DNA is generally present in the chromosomes of the cell nucleus, whereas RNA is present in the cell cytoplasm. The nucleotides are in living cells, associated with each other and with proteins to form nucleoproteins. Nucleoproteins are present in all living bodies down to the simplest viruses.

The nucleic acids are the ultimate self-replicating and mutable substances of living cells. They form the genes, the bearers of inheritance from parents to offspring in all species. This inheritance consists of information required for the manufacture of protein molecules. The protein molecules are unique for each species. The nucleic acids and proteins thus form an interlocking and interdependent system. This arrangement is unique for each individual and each species. Change in this interdependance produce the variation in the individuals of each species as well as the marked differences between various species of organisms.

The building blocks of the nucleic acids are the nucleotides of which there are four kinds in DNA. Genetic information is encoded in linear sequences of these nucleotides which are located on the long DNA chains. The same basic nucleotides are found in all known species. Three successive nucleotides on the DNA form one gene. The number and sequence of genes of a man differ from those of any other animal or insect only in the number and linear arrangements of nucleotides on DNA or genes.

Now the dependence of gene protein relationship is clear. Genes are innumerable and the number of proteins are also innumerable. But the building blocks of genes are only four nucleotides and the building blocks of proteins are only 20 amino acids. Inheritance consists of transmitting genes or nucleotide sequences on DNA molecules from parents to offspring. Here transmission is not sufficient. These coded messages of nucleotide sequences should be translated into the 20 amino acids which are the basic building blocks of animal bodies. The number and sequence of amino acids determine the specific protein or enzyme required for the formation and facilitators of body chemistry to function. So the Phenotypic expression and its functioning of any organism depends on its genotypic inheritance. This works through the translation of the 4 nucleotide sequences into the 20 amino acid sequences. This is the genetic code. This is the same for all organisms. The nucleotides are the same, the amino acids are the same and the genetic code is also the same. Thus all organisms are inter-related. If life originated once on our Planet Earth, it is sufficient. Mutation and natural selection would do the rest to produce all the other organisms. The nucleic acids and the proteins of all species are built out of the same building blocks according to the genetic code of 4 alphabets producing 20 words which is the same for all living species. It is their sequence and number which makes all the difference between species.

Even the biochemical route did not take us to the origin of life. Nevertheless it has thrown a lot of light on the possible organisms which might be the first. Molecular biologists are spending so much time studying RNA because they believe its roots reach back close to the dawn of life. Before DNA and proteins evolved, before even the simplest bacteria appeared, life may have consisted solely of RNA molecules floating in the primordial sea, replicating, mutating and undergoing natural selection on their own. As DNA and proteins evolved, these more specialised molecules took over most of the work from RNA. In due course of time these separate molecules found a home inside a protective membrane forming the very first cells. From these cells arose the three main branches of life bacteria, archaea and eukaryotes.

We have bacteria, archaea and eukaryotes existing now independently. But we do not have any precursor units from which they evolved in independent existence. The highly complex nucleoproteins and even their constituent parts the amino acids and nucleic acids are found only in living organisms. We can not synthesize them in laboratory experiments. This means only one thing. The conditions under which the primordial life arose do not exist now. Also we are unable to produce the same conditions in laboratory experiments. However scientists tried to visualise the conditions under which the constituent parts of the self-replicating molecules could have arisen. It means they should visualise conditions prevailing on the Earth some 4 billion years ago when it was cooling down to a temperature range which could sustain life.

Such a free living self-replicating unit might be regarded as a single gene. Mutation could then lead to formation of gene aggregates. Such gene aggregates could be regarded as independently existing chromosomes. Further mutations might then lead to accumulation of other metabolites around the chromosomes and the complex so produced could be regarded as an exposed nucleus. Bacteria in general are regarded as representing this level of organization. Next cytoplasm might be acquired but not separated from the nuclear material, as in the blue green algae, or in the larger bacteria. Finally mutation could result in the formation of typical cells with nucleus and cytoplasm separated by a membrane. Within such cells lie all the potentialities of higher organisms. It has taken from the first self replicating molecules to the formation of single celled organisms with cytoplasm and nucleus separated, nearly one billion years. The possible details of such a chemical evolution of most elementary forms of life have been worked out by the Russian scientist Oparin.

The above scenario is mostly conjectural. We have to see in laboratory experiments whether it is possible, particularly whether self-replicating molecules could be produced under conditions prevailing on the Earth some 4 billion years ago.

So what is the kind of atmosphere that prevailed on the Earth that helped the origin of life. The Earth's atmosphere is of secondary origin. That is, the Earth was formed out of solid materials which had no atmosphere associated with them. The atoms that formed the atmosphere later were combined in solid chemical compounds first. These were later disintegrated by the energy of the gravitational force, the radioactivity of the earth and other forces from the solar radiation etc. These liberated gases formed the Earth's atmosphere and Oceans. This secondary atmosphere was reducing in nature. A reducing atmosphere would contain methane, ammonia, water vapour and hydrogen.

Experiments were conducted in 1952 by Stanley Miller, then a graduate student of the Chemist Harold Urey at the University of Chicago. Using only the most abundant gases that were present on the early Earth such as water vapour, ammonia, methane, hydrogen sulphide etc., and passing an electric spark through these gases, the essential building blocks of life, the constituent parts of proteins and nucleic acids were produced.

The molecular building blocks must be put together in the correct sequence. Life is certainly more than the amino acids that make up its proteins and the nucleotides that make up its nucleic acids. But in ordering these building blocks into long chain molecules, there has been substantial laboratory progress. Amino acids have been assembled in primitive earth conditions into molecules resembling proteins. Some of them feebly control useful chemical reactions as enzymes do. Nucleotides have been put together into strands of nucleic acids a few dozen units long. Under the right conditions in the test tube, short nucleic acids can synthesize identical copies of themselves.

The smallest living things known today are the viroids. They are composed of less than 10,000 atoms. Viroids are composed exclusively of nucleic acids unlike the viruses, which also have a protein coat. However the viroids and viruses are now parasites. They need the cells of host body to replicate. The first living things may have been some thing like the free living viroids only a few hundred nucleotides long. That would have been a piece of RNA which is the subject of intensive study by the scientists now to know the secrets of life. How RNA came to be remains a mystery giving scope to many theories. More than 3 billion years ago probably the first cells formed. They were probably much like the modern Prokaryotes — bacteria and archaea.

They are now found in the mineral rich hot geysers that well up from the interior of the Earth in the Yellow stone national park in U.S.A. These seem to be the most primitive living things available to us now. These could live only near the mineral rich volcanic geysers. They contain no nucleus in their cells.

The next development came some 2 billion years ago. Some of these ancient cells could make their own food by a process known as photosynthesis. This requires Hydrogen and Hydrogen is obtained from water, a plentyful source. Those which were able to exploit this new source of food were no doubt very successful. This advantage came to them through mutation and natural selection. What we now call Darwinian Process of evolution. These are called blue green algae. These are several species and now collectively called cynobacteria. These also lack a nucleus in their cells. They reproduce by cell division like other bacteria. Blue greens flourish wherever there is moisture.

When hydrogen is removed from water, oxygen remained. Blue greens removed hydrogen from water for their photosynthetic activity. This happened for many millions of years and all these years oxygen accumulated in the atmosphere. This resulted in two far reaching developments. First is that the atmosphere became oxygen rich. Ozone which is oxygen with three atoms formed the upper layer of the atmosphere and this cuts out the ultraviolet rays which are lethal to higher life forms. This also means that the source of energy which helped the synthesis of amino acids and sugars from the primordial seas was no longer available. Hence origin of life again on the Earth is not possible. The other development is that the oxygen rich atmosphere permits the evolution of larger animals.

The initial Prokaryotes contained only genetic materials and ribosomes where the cells produce their proteins. By about 2 billion years ago we have bacteria which contained chloroplasts also which allowed the bacteria to manufacture its own food using the energy of the sunlight and carbondioxide and hydrogen. The removal of hydrozen from water produced an oxygen rich atmosphere. By 1.2 billion years the bacteria like cells evolved further by capturing other cells we now call mitochondria. These mitochondria are power houses which help in the metabolism of the cell in breaking down the carbohydrates and produce energy for the cell for its other activities. By this time the cell also developed true nucleus wherein lies its genetic material. The nucleus is covered by a membrane. Between it and the outer semipermeable membrane is the cytoplasm wherein lies the important mitochondria, chloroplasts, ribosomes and other important functional parts of the cell. Here is a full cell from which all other higher organisms developed. These are known as Eukaryotes. First the cell reproduced by simple division. This did not produce much variation for the natural selection and evolution to operate. Gradually the cells began to exchange genetic material with other cells and a sort of primitive sex developed. This produced much variation and a profusion of organisms developed.

Australian geologists unearthed fossils of some multicelled organisms at a site called Ediacara in 1940s. Before the Ediacaran species appeared some 600 million years ago, evolution worked mostly on the microscopic scale. Bacteria and other micro-organisms appeared between 4 billion and 3.5 billion years ago. Algae emerged more than 1.5 billion years ago. In the next billion years single celled forerunners of animals appeared. Throughout this vast stretch of time, life remained small and simple, kept in check by lack of Oxygen. The oceans and atmosphere held such scant quantities of this gas that they could not support the evolution of large organisms. The blue green algae took that long to fill the atmosphere and the Ocean with Oxygen.

The long Pre-Cambrian period which lasted between 4000 million to 540 million years ago ended. By 540 million years ago life suddenly became explosive. The calm of the Pre-Cambrian was disturbed by the evolution of many forms of life forms, some of them are predatory. Some ascribe this sudden change to a change in sea water chemistry. Others ascribe this to appearance of new mutations. Both causes might have contributed to the cambrian explosion, but Oxygen was abundant now giving opportunities for larger life forms to evolve.

We have already noted that the origin of life is still an elusive matter. We merely noted that it might have taken place as long ago as 4000 million years ago. The conditions existing on Earth at that remote past do not exist on Earth now. Scientists are also not able to produce anything like a self replicating and mutating molecule in the laboratory test tube so far. This fact was seized upon by other scientists to argue that life might have originated somewhere out there in the solar system or even in the vast reaches of the Universe and ferried on to Earth on the asteroids and comets which bombarded the Earth till 3.8 billion years ago. Even today a constant rain of microscopic interplanetary dust particles delivers the same kind of compounds which might have contributed to the primordial soup and the living things that arose from it. Hale-Bopp Comet which came close to the Earth in 1997 revealed that there are materials in its composition which are prerequisites of amino acids and nucleotides.

This extraterrestrial origin of life is known as panspermia theory. It was first proposed by Arrehenius, a Dutch chemist in the last years of 19th century. Its recent proponents in 20th Century are the British Astro-Physicist Fred Hoyle and his associate Chandra Vikramasinghe. Francis Crick the British scientist who decoded DNA structure along with the American James Watson also proposed a form of panspermia theory. The American Space Agency NASA had plans to bring samples of micro-organisms from Mars if they do exist there. The most interesting question such micro-organisms from Mars or any other extra terrestrial source should answer is whether they obey the laws of genetic code which is common to all organisms on planet earth. If so the genetic code would be universal not merely Earth bound.

This has important philosophical implications. Scientists are fairly unanimous in their view that the Physical and

Chemical laws are applicable throughout the universe. They are unable to ascertain with the same authority and confidence that the biological laws are also the same throughout the Universe. That is, the sequences of the Nucleotide pairs A-T, T-A, G-C, C-G in triplets should code for the same amino acids. If this genetic code holds good for the organisms from outer space then we could ascertain that the biological laws are also Universal.

5

THE PROCESS OF EVOLUTION

The previous chapter on 'The Origin of Life' and this chapter has much in common. They deal with practically the same subject but the emphasis and details are different. Both took a scientific turn starting from the middle of the 19th century. Earlier views were mostly theological and religious views. As Copernicus, Kepler and Galileo faced opposition from religious authorities, so did the 19th century biologists and naturalists faced opposition from religious authorities as well as the general public at large who were believers in the story of Genesis in the Holy Bible. Charles Darwin was particularly opposed for his views on Organic Evolution set out in his book 'On the Origin of Species'. Even now some religious societies are burning midnight oil to debunk Darwin's theory of Evolution. The Indian religious Pundits however have no such opposition to Darwin's theory. They contend gratuitously that the avatars of the Hindu God Vishnu are nothing but Evolution of lower forms of life to higher forms.

By Darwin's time nearly three centuries of knowledge about the flora and fauna of the world was available to the sea faring nations of Europe. The amazing diversity of the life forms and the varied circumstances and living conditions they occupied were sought to be explained by a theological argument and by a naturalistic argument. The theological argument is that the Creator has endowed every living species with the kind of features which enabled that species to live in its habitat. Otherwise species were created once for all with their fixed features.

The naturalists were not satisfied with the theological argument. They sought a rationalistic argument. For this purpose they gathered ever more information about the flora and fauna of the world. Charles Darwin (1809-1882) an English naturalist went on such an expedition on the ship 'Beagle' to South America and brought much material with him. Working on it for many years he finally published his great work 'On the Origin of Species' in 1859.

Darwin theorized that species evolved one from the other and filled up all the habitable parts of the Earth over millions of years. His contemporary the geologist Lyell theorized that the surface of the Earth was not the same throughout its existence but was changing by the forces exerted by Volcanoes, earthquakes which raised the land masses and the forces of wind and water which eroded the mountains to create the soils and plains and sands of the beaches. The Earth did not remain the same through the many millions of years of its existence. The conditions varied due to natural forces.

So the organisms also have to vary to cope up with the ever changing external conditions. This variation may not be sudden but ever so slowly both varied, the external conditions and the organisms which lived in these ecological niches. The organisms had to evolve from pre existing organisms and were never created readymade. Thus natural forces produced a variety of habitats or 'ecological niches' which were occupied by the millions of species of living things both plant and animal.

Earlier naturalists, particularly the French naturalist Lamark (1744-1829) thought that evolutionary change occurred due to environmental factors. Then they were passed on to the offspring. According to this theory a body builder acquiring huge muscles could pass them on to his children. But this was not so. Darwin guessed it correctly, that it was inherited characters that were passed on to the offspring if the organism lived long enough to produce offspring. The organism should acquire the characteristics at the time of its conception itself by its mating parents and not after birth. If the organism lived long enough *i.e.,* successful in its habitat to reproduce, then it could pass on its slightly varied characteristics to its offspring. Thus species produced new species over many generations to adapt themselves to the varying habitats of the world which never remained constant.

Darwin's theory of evolution rests on three observed facts and two deductions from them. The first fact is that all living organisms both plant and animal reproduce much in excess than actually survive. A salmon produces 28 million eggs and an oyster produces 114 million eggs in a season. Think of what would happen to the Earth if all those eggs were to grow to adult size and reproduce again. In Darwin's days the slowly reproducing humans doubled in numbers. Darwin calculated that in another 1000 years there won't be even space to stand on Earth if multiplication of human race were to continue at the same rate. The second fact observed by Darwin is that populations of species tended to be more or less constant. From these two observed facts Darwin drew his first deduction. It was, that there has to be a struggle for existence among the members of the same species as well as between species for the available food and space. The struggle is not tooth and claw struggle. Predators, parasites and lack of food and adverse atmosphere often keep the organisms in check. Many will become food for others. Only man does not. Thomas Malthus (1766-1834) in his essay on the Principles of Population noted that only famine, disease and war kept the world's population of humans in check. When once these are kept in control humans managed to increase to 6 billions by the end of 20th century. We will discuss the implications elsewhere.

The third fact observed by Darwin is that the members of a species vary from each other. In mankind we see this variation clearly. No two individuals are similar. Even children of the same parents are not similar. They vary from each other. Now Darwin made his crucial second deduction. He pointed out that some of these variations are favourable and some not for that organism to survive. Those variations that do not favour the organism to survive are selected away and only those variations that favour the organism to survive will be preserved and passed on to the next generation. Thus nature itself acts as a sieve to select the favourable variations among the members of a species. This process is known as natural selection and is at the centre of species formation and organic evolution.

We have seen that the inheritable characters are the ones that are selected if they help the organism in survival. We now call these characters genes. The sum total of genes of an organism is called its 'genotype'. Their expression in physical characteristics both external and internal is called 'Phenotype'. Now the physical conditions present some variation in every season. There may be more rain or drought, more cold or more heat, more wind or less etc. The survivors of each season are likely to differ from those that do not survive in their genotype by giving them a better fitness. This results in automatic natural selection which acts as a sieve on the genotypes.

Thus natural selection, struggle for existence and survival of the fittest have a meaning in the natural world which is entirely different from human affairs. Unfortunately some sociologists, anthropologists and economists at the end of 19th century and at the beginning of 20th century used these words in the setting of human societies and tried to justify their claims for the superiority of some races over others, for opposing poverty alleviation measures and for supporting *Laissez faire* economics. Thus science was misused by those who have no training in scientific matters like Bishop Wilberforce of Oxford who tried to denounce Darwin.

Thus 'struggle' 'competition' survival of the fittest have different meaning when applied to the process of 'natural selection'. Thus trees 'struggle' against wind by developing stronger root system. Struggle against drought and dryness by developing cactus shapes without leaves. Animals and birds struggle against cold by developing fur and feathers for heat insulation and temperature regulation mechanisms. Insects struggle against predators by developing mimicry or unpalatable tastes and odours. All creatures struggle and 'compete' for food and mates and avoid predators.

But this does not always produce fights with each other. The weaker one often backs off and become less successful in survival and may get selected out. Its genes will not be passed to the next generation. Only the successful one passes on its genes to the next generation. Thus natural selection ensures that the best possible genotypes will survive in successive generations. In Darwin's days it was not understood how the variations in the members of a single species came about, though it was understood some variations would be selected by nature to be passed on to the next generation. That is how evolution functioned.

At the same time Darwin was formulating his ideas in 'The Origin of Species' an Austrian monk Gregor Johnn Mendel started some experiments in 1856 on garden peas to find out what factors contributed to determine inheritance. He worked in a small town, Brunn now Brno in Czeck republic. He reported his results to the Brunn Society of Natural Science and his monographs lay unnoticed for nearly 40 years until the end of 19th century when they were simultaneously discovered by three scientists.

Mendel selected seven easily identifiable physical characteristics of garden peas for his experiments. They are round or wrinkled seeds, yellow or green seed interior, grey or white seed coats, inflated or pinched ripe pods, green or yellow unripe pods, axial or terminal flowers, long or short stems.

Mendel produced hybrids of each characteristic with its opposite number thus crossing round seeds with wrinkled seeds etc., for the seven characteristics he chose. When the seeds were ripe Mendel found only round seeds in the hybrids between round and wrinkled seeds. It was so in the case of the other characteristics also. Only the first out of the pairs appeared in the hybrids. Mendel chose to call them the dominant character and the other one a recessive character.

Mendel next planted these hybrid seeds and allowed them to fertilize themselves in the natural way of the peas. When opened the ripe pods he found both round and wrinkled seeds inside the same pod. The lost wrinkling of the first generation reappeared again in the second generation. The ratio of round to wrinkled seeds was 3:1. It was the same with the other characteristics also.

Mendel planted again the 3:1 second generation seeds to see what will happen in the third generation. The wrinkled seeds produced only wrinkled seeds though Mendel planted them as many as for seven generations. But it was not so with the round seeds.

When he planted them for the third generation two round seeds produced both round and wrinkled seeds in 3 : 1 ratio, one round seed produced only round seeds. Though all the three round seeds of the second generation looked alike, two of them are hybrids between round and wrinkled. Only one was true round. The wrinkledness was a recessive character and was hidden in the germ plasm of the seed. Only the round quality which was dominant appeared in the outward physical characteristic or phenotype.

Mendel called the dominant characteristic 'A' and the recessive characteristic 'a'. When 'A' and 'A' came together it means two dominants and a round Pea. When 'a' and 'a' combined it meant two recessives and a wrinkled pea. It was only when 'A' the dominant round combined with 'a' the recessive that hybrids occur.

If each trait marked a separate hereditary factor, then he was obtaining every combination that could be formed. Combine 'A' and 'a' and only one unit could be formed 'Aa'. But if 'Aa' and 'Aa' came together three different combinations could be made 'AA', 'Aa' and 'aa'. Thus with one pair of hybrids three kinds of offspring would be produced; with two pairs of hybrids, nine kinds; with three, 27. The combinations would pile up three times, three times three in cubic power.

Mendel's laws could be summarised: (1) Heredity is determined by a large number of discreet units which are independent of each other (2) When each parent contributes the same kind of units then a constant character shows in the phenotype. When each parent contributes a different unit, then a hybrid results. Its Physical appearance depends on the dominant unit. The recessive unit lays dormant in the germ plasm of the hybrid. When the hybrid forms its own reproductive cells, the different units segregate. The inheritable units are not like milk and water mixing together undistinguishably. They are like beads of different colors. They separate and maintain their individuality. These units are much latter called genes.

In Mendel's days microscopic techniques were lacking to look into the cells of his beans and look for these inheritable units which maintain their independence throughout the combination and segregation of the genetic materials of plants as well as animals.

Hugo de Vries was the botanist at Amsterdam University in the latter part of 19th century. He was observing the plants called evening primroses. They presented the usual variations among individuals of the same species. Next year however he found ten plants of a different variety of the primroses. They have their petals differently shaped. During the next twenty years de Vries observed more than 50 thousand evening primrose plants and found no less than 8 different species. These differences appeared suddenly and not by gradual degrees. In one species the veins were red. He concluded that each of the characters of the species were produced by separate independent hereditary units. If there is a sudden shift in that hereditary unit that is called a mutation by de Vries. These hereditary units are now called genes.

Natural selection will operate on such mutations after they occur and express themselves in physical form. They will be selected out if such mutations are not useful and they will be preserved if such mutations are useful. Thus de Vries formulated the important concept of 'mutation'. In the process he also discovered Mendel's original work on garden peas and rules of heredity. In the same spring of 1900 another German and an Austrian scientist also 'discovered' Mendel's work and confirmed it. Thus Mendel's laws of heredity were established as the basis of heredity. Something physical has to pass from parents to offspring. What is it and where it lies. In the days of Mendel microscopic studies of cells were not available. By the end of 1900 such studies were available.

All living bodies consist of cells. There may be different cells like bone cells, muscle cells, blood cells, nerve cells etc. But basically it is the cells that constitute the body of any living organism. The cells continuously divide and form other cells which will give growth to the organism and replace worn out cells. The human body starts with a single fertilized egg cell called zygote and by the time of birth consist some 200 billion cells.

The cell consists of a nucleus enclosed in a membrane. Outside it there is colloidal solution called cytoplasm again enclosed in a semipermeable membrane, through which metabolic activity occur. The cells divide in two ways. One is the division of body cells and the other is the division of reproductive cells. The nucleus of the cell contains some thread like bodies and some proteins covering them. These thread like structures occur in pairs and the number of pairs are characteristic of that species. Thus humans have 23 pairs. Mendel's peas seven pairs and the common fruit fly Drasophila four pairs. These threads take the red dye and so named chromosomes or simply coloured bodies. When coloured they could be seen under a microscope. When a cell is about to divide they split and move to the opposite ends of the cell.

Their complementary bodies are formed out of the material available in the cell itself and the cell splits in half and a cell wall grows between the two halves and in humans in about 80 minutes there are two cells now instead of one. What is more the full chromosome compliment of 23 sets are also present in both. But when a new egg or sperm cell was to be formed, only half of each set of chromosomes enter the new sex cell. When the egg cell is fertilized by the sperm cell the full compliment of 23 pairs of chromosomes is assembled and the fertilized egg grows into an individual baby. This process is as true of humans as in the case of any other organism. This is a universal property of living matter. Only the number of pairs of chromosomes differ from species to species.

By 1900 when Mendel's work was rediscovered this mechanism of cell division with the division of chromosomes and their recombination in the fertilization of egg cells was well understood. Scientists wondered whether this process is not what carried Mendel's hereditary units. The chromosomes provided just such mechanism. Some of the experimenters found that some of the characteristic traits stayed together. The number of traits are more than the number of chromosomes. So when chromosomes separated it is but natural to expect that all the traits on that chromosome stay together.

Thomas Hunt Morgan of Columbia University in New York set out to find out. After nearly 20 years of work he did find out some remarkable features of the chromosomes and its constituent bits which he named genes. The genes arrange themselves in linear order on the length of the chromosome and at specific locations. He chose the common fruit fly, Drosophila Melanogaster for his experiments because it develops from egg to adult size in only 12 days, where as Mendel's peas require a whole growing season. Morgan's experiments required study on a large number of individuals and quite fast. Besides the chromosomes in the salivary glands of Drosophila are particularly lengthy which suits well Morgan's purpose of observing them under a microscope.

However Morgan needed mutants which could be tagged as controls. Here crossing and recrossing of wild flies won't do. After an year of patient breeding of the flies on banana mash in milk bottles he obtained a mutant fly. It was a white eyed male. The normal eye colour of a fly is red. He mated it with a red eyed female. He obtained the first generation all with the dominant red eyes as per Mendel's laws. They are all hybrids. When the hybrids are mated with each other, the second generation produced red eyed females, red eyed males and only white eyed males in the proportions as per Mendel's law. There was not one white eyed female. This at once suggested that the hereditary unit for producing white eyes was on the chromosome which also carried the hereditary units which produced male sex. So these two characteristics were linked together.

This suggested to Morgan that other characteristics of the fly and its mutations could be located on its chromosomes and could be mapped. This is a tedious and lengthy process since mutations that could be clearly identified and used as tags are hard to come by. Nevertheless he and his enthusiastic team of students and associates found such groups of characters which appeared together. There were four such groups. There were two large groups which contained as many as 50 characteristics like body colour, shape, wing colour and shape, eye colour and shape bristle size and shape etc. There was one tiny group and another large group which was associated with the sex of the flies.

Microscopic examinations of the chromosomes and their division and reunion agreed well with the experimental observations as to the size of the chromosomes. There were two large pairs and a tiny pair. In the female there was another large pair. In the male also there was another large pair but one of the pair was hooked at one end as opposed to the two in the female which are both straight. Morgan identified them as the chromosomes responsible for the sex of the flies. He named the two straight chromosomes in the female and the one straight chromosome in the male as X chromosomes and the one hooked chromosome in the male as Y chromosome.

We have seen that in the formation of the reproductive cells only one chromosome of each pair entered the reproductive cell. Thus the female has two eggs both of which contain X chromosome along with one member of the other three pairs. Whereas the male sperm cells are of two types. One half of the sperm cells contain X chromosome along with the other three and the other half of the sperm cells contain Y chromosome along with the other three. Now after crossing and fertilization of the egg cells with sperm cells only two types of combinations are possible. An XX or XY. The first combination results in female offspring and the second combination results in male offspring. When sufficiently large numbers are taken the ratio of females to males should be equal. This is as true of fruit flies as of humans or any other bisexually reproducing organism.

However some human societies resort to queer ways which change this natural balance. Many social groups in many countries including India resort to the wholly unacceptable practice of infanticide and foeticide of females. This is reducing the number of females per 1000 males. As per the census of 2001 in India there are only 923 females per 1000 males. This ratio is even adverse in the states of Punjab (793/1000), Haryana (820/1000) and Tamilnadu (939/1000). Sociologists are already ringing alarm bells as to the possible consequences of such adverse ratio between the sexes.

As already noted there is only one type of egg in the female carrying X chromosome only. Whereas there are two types of sperm cells in males approximately half carrying X chromosome and the other half carrying Y chromosome. If the X carrying sperm cell fertilizes the egg then an XX will result and a female baby is born. If Y carrying sperm fertilizers the egg then the full chromosome compliment will be XY and a male will result. So it is the male who is responsible for the sex of the child. Strangely in many societies it is the female who is blamed for the sex of the child. Discrimination against girl children is quite common in rural and poor families.

The work of Morgan and his associates established without any doubt that the location of the hereditary units is the chromosome inside cell nucleus that exist in every cell of every organism. He further showed that the bits of chromosome that are responsible for individual characters are linearly arranged on the chromosomes and he gave the name 'genes' to them. They have even mapped some gene locations on the chromosomes of drosophila. These bits of matter called genes are the only physical substance passed on from parent to offspring from the origin of life till now. However we do not yet know how life itself had originated. Nevertheless there is an unbroken continuity in life. In this sense all life is kith and kin.

Now the focus of attention is the gene and the mutations that occur in them. Morgan and his associates had no way of inducing artificial mutations. They had to breed literally thousands of fruit flies and examine them laboriously under magnifying glasses for any mutations. After 20 years labour and breeding 20 million fruit flies their count of mutations was only 400.

Then one of the associates of Morgan, Herman J. Muller wondered whether mutations could not be induced by artificial means. The gene was tough subject well protected inside the cell nucleus inside cytoplasm and cell wall inside the tissues of the body. He tried all sorts of gross physical and chemical methods and failed. Finally he hit upon the idea of using X-rays since X-rays are such minute yet powerful packets of energy it will act on a minute point of matter and avoid other points thousandths of a millimeter away. So he put thousands of fruit flies in gelatine capsules and irradiated them with not too lethal X-rays. The result was astonishing.

He obtained many mutant flies when he crossed the X-ray treated flies with non-treated flies. However many of the mutants were lethal. They died before they could produce offspring or became sterile.

Muller induced artificial mutations in fruit flies hitting them with X-rays. Others tried the same techniques on other plant and insect organisms and obtained radiation induced mutants. However many of them proved to be lethal. There is natural radiation in the atmosphere in the form of gamma rays and cosmic rays and radiation released from radioactive material. All of them produce random mutations in plants as well as animals and insects. Many of the mutations are useless and their carriers die before they could reproduce. What is more these mutations should be in the reproductive cells not in body cells to be of any use to the offspring.

In the second half of twentieth century starting from the first atomic explosion at Alamogordo, New Mexico state in U.S.A many atomic bombs were exploded. The radiation let loose into the atmosphere through these atom bombs and hydrogen bombs is posing a grave danger by its potential to produce lethal mutations in many life forms as well as humans that could seriously endanger life on earth. The accident in Chernobyl atomic power plant in Ukraine of the former U.S.S.R. is a poignant reminder that radioactive substances whether released in atomic explosions or its use for peaceful purposes poses a grave danger to mankind as well as all life forms.

As a result of all these cumulative explosions as well as accidental releases from nuclear reactors etc., man is exposed now to a higher degree of radiation than half a century ago before the explosion of atomic devices. The damage done to the germ cells is proportional to the amount of radiation exposure and each explosion increases this amount of radiation in atmosphere. It comes back to the earth in the form of rain or snow or other forms of precipitation. It effects the germ cells cumulatively and they transmit their accumulated changes from generation to generation. Most of them are lethal and recessive. The danger is these recessives live on even though the individual dies. The advances in modern medicine are taking care to see that the carriers of these deleterious genetic endowment are not selected away as in wild animals or primitive people who have no benefit of medical care. Consequently human gene pool is getting degenerated by carrying all this genetic load of lethal and recessive genes. This is one grave danger mankind has to set right. We will see latter if mapping of genes of man will be of any help.

The work of Morgan and Muller had shown that the seat of heredity is the chromosomes within the cell nucleus and more particularly to sections of chromosomes given the name genes and are arranged in linear order on the chromosomes. They have even mapped the specific points on the chromosomes which corresponded and gave rise to specific bodily and functional features to the four pairs of the chromosomes of fruit fly Drosophila. Some of the features identified by Morgan and his team are white eyes, black body, vestigial wings, bristle shape and size etc. They have nearly identified some four hundred factors like this to specific locations on the chromosomes.

Beadle and Tatum had taken up the genetics work at California Institute of Technology (Caltech) in the early thirties of 20th century. They chose the red bread moulds called Newrospora for their work. Their objective was to establish the gene as the controller of chemical reactions in the cell and thereby of the whole body of the organism. Since every cell in the body of the same species of organisms contains the same set of chromosomes, Beadle and Tatum's work established convincingly that genes alter the chemistry and hence the whole functioning of the organism. A change in one gene-one section of a chromosome modified one reaction. This has affected the way that the organism produced some materials essential for their lives. The red bread mould Newrospora produced a mutant when irradiated by X-rays which was unable to produce the Vitamin B-6. In due course they produced many more mutants and found out what was the substance they were unable to produce which was needed for the organism to survive. Thus genes exerted their control by altering the chemical reactions which effected the organism. Genes thus are not only instrumental in producing new species but also the controllers of their mechanism.

Further progress in genetics required delving deeper into the gene and its chemical properties and structure. A Swiss

Chemist named Friedrich Miescher discovered in 1869, what is now called deoxyribonucleic acid, trying to break down some cells. The cells disintegrated under the treatment of strong acids but some material of the nucleus remained intact. It is different from all other known cellular material. Other scientists found that it is an acid and its structure was thread like and its molecules were huge. It occurred only in chromosomes inside the nucleus of the cell. The powder was named DNA for short. Later a second nucleic acid was found and named Ribonucleic acid or RNA for short. It is found in the cytoplasm of the cell, that viscous and transparent substance between the cell wall and the nucleus.

M. Hershey and Martha Chase working at the Biological Laboratory, Cold Spring Harbor, New York, were conducting experiments on bacteriophages in the early 1950s. Bacteriophages or simply the phage is a length of chromosome or a length of DNA with a protein coat. They are viruses which infect the bacteria. The two experimenters injected the viral DNA of the phage into the host bacterium in an effort to infect the bacterium. The protein coat was left outside. In 24 minutes the DNA which entered the bacterium produced copies of itself as well as the protein overcoat breaking the contents of the host bacterium. The DNA which entered the bacterium contained all the instructions to produce more DNA with the protein over coats. These experiments proved that the DNA or its close copy RNA are the sources of heredity.

All DNA was made of same materials, some five sided sugar molecules joined by a phosphate molecule and four nucleotide bases called adenine, thiamine, cytocine and guanine known as A, T, C and G for short. Thus there are six kinds of pieces in the DNA, whether the DNA was extracted from liver cells, blood cells or of any body part. What is more the DNA also contained the same six components whether it came from humans, animals, fish or plants.

The DNA is universally the same. Then the question was how this molecule in the cell which is too small to be detected except under the great enlarging power of electron microscope could contain all the instructions of the millions of varied organisms both plant and animal. Therefore the secret of the creative diversity of DNA has to be sought in its structure and not in its component parts. The way the six parts were arranged in the DNA has to account for the millions of forms it produces.

In the early 1950s Francis H.C. Crick of England and James W. Watson of America teamed together and started building wire models of DNA molecule at the Cavendish laboratroy of Cambridge University. The six pieces that compose the DNA molecule were well studied and their shapes and measurements were known. The five sided sugar molecules were joined together by a phosphate molecule into long chains of alternate sugar phosphate molecules. The four bases A, T, C and G were attached at right angles to the sugars.

A second chain was also formed. It was turned upside down and brought closer to the other and the two were twisted together. The bases with their hydrogen atoms acting as joiners fused together. It is like a spiral staircase, a double helix. The sugar phosphate chains formed the frame of the staircase and the bases formed the steps or rungs of the staircase. X-ray studies of DNA showed that the X-ray pattern was identical for DNA from what ever organism or body cells it was extracted. If the DNA structure of the various organisms was similar how the millions of species and further the countless individuals of each species could be accounted for.

That drew the attention to the bases A, T, C and G. They are called bases because they could combine with a hydrogen ion. Experiments have previously established that there are equal number of As and Ts and again equal number of Gs and Cs. This suggested that the As joined with Ts and Gs with Cs. The As and Gs are long and Cs and Ts are short. However each pair A-T, and T-A, C-G and G-C are equal in length to each other and they perfectly fitted the sugar phosphate frame. The four base pairs could follow each other in any order. This accounted for all the differences in the millions of organisms and species.

In a single DNA thread there might be ten thousand steps. In the forty six chromosomes of human body there might be a hundred million steps. The varying arrangements of these millions of steps could produce the 6 billion human beings that inhabit the earth and no two of them are identical except identical twins. The same also account for the various species and the variation in each of their members.

All living matter duplicate itself. This happens through cell division. During cell division the DNA coil unwinds itself vertically breaking the hydrogen bonds that join two nucleotides that form each step. In the cell nucleus and cytoplasm there are building blocks available that are continuously being supplied to the cell by the metabolism of the food that the organism consumes. With these building blocks a complementary coil to each of the divided half coils are built up and they join up with the half coils to form a full coil of the DNA. In the place of one DNA coil there are now two DNA coils exactly like the one which had separated into two halves before. This DNA is not any DNA but characteristic of that species and animal concerned. This is the process of DNA duplication in cell division that is a continuous process in every organism.

There are only the four different steps in the DNA that are the variables. These four bases produce the twenty different amino acids by their different arrangement. The twenty different amino acids by their different sequencing and number in each sequence produce the more than hundred thousand proteins that are the building blocks which is responsible for the limitless variety of the living world as well as the controllers of its mechanism.

Francis Crick also solved the problem of how the four bases of the DNA produce only 20 different amino acids. Three steps code for one amino acid. When this is known it is only a matter of simple mathematical calculation of permutations and combinations to see how many groups of three units could be made out of four different units. It is sixty four. But there could be duplications in these sixty four. Separating the duplicates we have only twenty distinctively made up groups of three units each. So we have only twenty distinctive and separate amino acids.

The genes an organism acquires from its parents, half from each, is called its genotype. They are linearly arranged in DNA strands and covered by protein coats. Then they are called chromosomes. The chromosomes are generally arranged in pairs. The number of chromosomes and the number of genes therein are characteristic of that species and organism. Humans have 23 pairs of chromosomes and some 30,000 genes according to the latest estimates by the two groups involved in mapping the human genotype. These genes control the external appearance, colour, shape, internal organisation and its working in the minute details. This is called the Phenotype of the organism. The sequence of bases or steps in the DNA helix determine the sequence of amino acids formed and consequently the specific proteins and enzymes which are the ultimate building blocks and controllers of all the bodily chemistry that sustains the organisms.

If the sequence of amino acids is changed, a different protein is produced. The Phenotypic expression of proteins depends on the line up of steps in the DNA chain. Even change of one step could produce a different protein. In the disease of sickle cell anaemia it was found out that the haemoglobin molecules are controlled by a sequence where one gene is C-G, T-A and G-C. In the diseased person the T-A step is replaced by another G-C step. The change of one step resulted in the production of a different protein and the disease. Hence change in one step often leads to lethal mutations. Very often possessors of such lethal mutants die before they could reproduce and pass on the mutant genes to their offspring. Very often the mutants also remain as recessives and do not express themselves in the Phenotype.

When the outside living conditions change and two recessives come together and produce offspring then the recessive gene will express itself in the Phenotype. In the changed living conditions such recessive genes sometimes become useful and become dominant. The accumulation of many such mutations over time produce evolution through natural selection. Any such selection should help the organism adapt better to obtain food, evade predators, attract mates and leave offspring. Such favourable mutations spread through the deme *i.e.*, the inter breeding group quite fast. Statisticians like Fisher, Haldane, Sewal Wright etc., have worked out the rate of spread of such favourable mutations and in how many generations a new species could be formed out of the old one.

This is the mechanism of natural selection which results in evolution and species formation. Acting over populations over many generations it changes organisms to adapt to the varying environments. Many types of ecological niches produce many species of organisms to take advantage of those niches. This mechanism is not understood in the days of Darwin, but we understand it now fairly well and nearly 140 years of research lay behind it since the days of Mendel and his experiments with garden peas in 1860s to the present day when the whole of human genotype is being mapped and work has been almost completed. It is announced in April 2003 that the work is completed.

6

VERTEBRATE EVOLUTION

The long period of Precambrian life had come to an end about 600 million years ago. Previous life forms were microscopic and single celled. Multicellular organisms developed only since about 600 million years ago. The first 100 million years of this period is known as Cambrian after Cambria the ancient Roman name for Wales where the rocks and the fossils of that period were first discovered. Geologists have constructed a picture of the conditions prevailed in the Cambrian Period. Shallow seas occupied a large part of the Earth's surface. The land masses were low lying and occupied much smaller area than they do today. There was no plant life except for lichens and a few very primitive plants. There was no animal life on land. Whatever life is there it was only aquatic.

Biologists classify animals into large groups known as 'Phyla'. There are 22 such 'Phyla' which are then sub-divided into classes, orders, families, genera and finally individual species. There are three important Phyla which are of interest to us. First the 'Chordates', which include the fish, amphibians, reptiles, birds and mammals. This is the Phylum to which we humans belong and by far the most interesting in a human view point and we will study its evolution in detail.

Between the sponges and corals which are the primitive multicelled organisms and the early fish which belong to 'Phylum Chordata', there are a couple of other Phylums which are of importance and which have evolved in the Cambrian seas and have contributed some physical features to chordates. The Phylum 'molluscs' include clams, oysters and scallops, snails and conches, shell fish and squids, octopuses and nautiloids. The Phylum 'Arthropods' constitute by far the largest group. Containing four major classes : lobsters and crabs, centipedes and millipedes, spiders and scorpions and insects.

Trilobites as its name implies had a three segmented body and an early Arthropod. It dominated the Cambrian seas and shared the seas with some early forms of molluscs. Trilobites developed a wide variety of shapes and sizes and quite advanced in their day. They lived a long period. They lasted more than 300 million years.

The Cambrian period lasted in this placid and largely uneventful way for a hundred million years. It was followed by Ordovician (500-425 million) named after an ancient British tribe in whose territory rocks bearing the fossils of this period were first located. This period was marked by much geological disturbance in the Western hemisphere. The oceans crept in for a time and divided North America into a group of islands.

A huge mountain chain reared on the east coast from Newfound land to Alabama. Many volcanoes erupted.

In the seas trilobites continued in enormous groups but their pre-eminence was threatened by the emergence of a number of molluscs. The Ordovician Period is also marked by the appearance of the first chordate. This was a pioneer among the back boned animals. It belonged to the jawless variety of primitive fish. The modern variety of this kind of fish is represented by the lamprey, which is a parasite on other fish.

The next period was named Silurian (425-405 million) after another British tribe in whose territory fossil bearing rocks of this period were first found. This period lasted for 20 million years and the general conditions were not much different from those of Cambrian Period. Low shallow continents, placid and warm seas prevailed. Eastern United States was transformed by coastal uplifts which created a land locked sea which evaporated to create a vast salt desert.

Life proliferated in the warm seas. The trilobites continued but in a variety of shapes and sizes. They have evolved protective armour. They are no longer the dominant creatures of the seas. The evolving fish became more mobile and rapacious to end the dominance of trilobites. This is also the age when small inconspicuous scorpions and millipedes ventured on to the land. Also some primitive plants crept on to the land. The land began to be populated finally.

The next age, the Devonian (405 to 345) was dominated by the advances made by the fishes. They are vertebrates and part of the Phylum 'Chordata'. They have developed into two major groups. One is the cartilagenous fish like the sharks and the other group is bony fishes. These ancient fishes developed into most of the modern fish species. These fishes developed true jaws, movable lower part with teeth. This is a major contribution of fishes to the vertebrate Phylum. All subsequent vertebrate classes developed upon this innovation to suit their special needs.

The group of fishes which developed jaws had two other evolutionary innovations to their credit. They also developed internal air bladders which could be used as lungs and stout muscular lobes on the fins. The Devonian Period was marked by shallow seas which came and went as a result of the slow undulations of the earth's crust. The air gulping and walking fish had certainly enjoyed an advantage in these shallow seas as well as the low lying mud flats in between. These have finally invaded the land masses at the close of the Devonian Period (405-345). This has ushered in a diversification of land living tetrapods in the next period called Carboniferous (345-280) and latter periods.

The Carboniferous age was so named because of the huge coal deposits that were formed during the crustal activity of this period. The fossils of this period which were mostly amphibian fossils were found in the coal mines. The land scape of this period was covered by huge fern trees and many kinds of evergreens. The crustal activity of this period together with the mountain ranges and volcanoes have repeatedly inundated these forests under sea. They were crushed by sediments and mud and much compressed into some of the world's great coal deposits.

Starting at the beginning of this period a particular group of fish which developed air bladders and stout muscular fins began clambering on to the land to exploit the rich insect food available in the forests and swamps. These have developed into a variety of amphibians. Their breathing and locomotion problems were partially solved by their fish ancestors. The land animals had to contend with greater gravitational pull. So these amphibians developed stronger backbones and stronger leg muscles. The backbone had two girdles, the thoracic and the pelvic and they were supported by strong bony legs and moved by strong muscles at a height from the ground. The leg bones were attached to the two girdles at right angles. So their gait was a bit awkward with a dragging tail behind. They have also developed good lungs for breathing. However they are not completely free from water. Their skin is permeable and their body fluids evaporate if they stayed too long out of water. So they had to frequently bathe themselves in water. Besides the problem of reproduction also tied them to water. They lay their fragile gelatinous eggs in water unprotected as most of the fishes do. The modern surviving amphibians, the frogs and salamanders do the same even today. The young of the frogs, tadpoles pass through free swimming aquatic stage before they can return to land in mature form. The extinct amphibians also did the same. Though their clambering on to the land was a great evolutionary advance, they still had to depend on water. This problem was solved by the reptiles which were dominant during the Mesozoic Era (230-63 million)

Geologists divide the last 600 million years of life on earth's history into 3 Eras. The first is Paleozoic era (ancient life), next Mesozoic era (middle life) and Ceinozoic era (recent life). The Paleozoic is sub-divided into six periods and lasted from 600 to 230 million years ago with a total duration of 370 million years. The Mesozoic had three periods and lasted for 167 million years from 230 to 63 million years ago. The Ceinozoic began 63 million years ago and is still going on.

We have so far dealt with five periods of the Paleozoic era. The last period the Permian (280-230 million) was the most inhospitable periods the Earth had ever witnessed. Mountain building increased and there was a drastic change in the climate. Severe ice ages covered parts of Africa, Australia and South America with glaciers. Other parts of the world turned to deserts. Drying seas produced great salt deposits in

Russia, Germany and United States. All these brought such a strain on many organisms, both plant and animal, they were unable to withstand the rigours of cold and extreme dryness. Many failed to survive. The trilobites finally became extinct after 300 million years of life in the seas.

However the reptiles which made their appearance first in the Carboniferous period (345-280 million) not only survived the Permian death but adapted well to the rigours of the period to become dominant creatures on land in the Mesozoic Era (230-63 million). These are the first backboned animals to completely free themselves from dependence on water. Amphibians must return to water to mate and lay their delicate eggs. The reptiles on the other hand developed tough cases for their eggs and they could lay them anywhere on land without fear of dehydration. The reptiles also developed tough skins and scales on their bodies that allowed them to wander on the land without periodically drenching themselves in water as the amphibians did.

The amniotic egg of the reptiles is internally fertilised, after which it is deposited on the ground or in some suitable place, or sometimes retained within the oviduct of the female until the young animal hatches. The egg contains a large yolk that furnishes nourishment for the developing embryo. It also contains two sacks, one being the amnion which is filled with a liquid and contains the embryo, the other the allantois, which receives the waste products produced by the embryo during the period of its hatching. All these are enclosed in a tough shell which protects the contents and allows gaseous diffusion of Carbondioxide outside and Oxygen inside to the developing embryo.

Once having become established through the perfection of the amniotic egg, and scaly tough skins which do not dehydrate them the reptiles quickly diverged in many directions and the main evolutionary lines that they were to follow in the Mesozoic Period were established. During the early Mesozoic Period reptiles flourished in a great variety of shapes and sizes, gradually getting bigger and bigger as time went on. Some of them returned to the sea. Others took to the swamps. By the middle of the era which is known as the Jurassic Period (181-135 million), they had evolved into the largest land animals the world has ever known. This was the age of dinosaurs.

Then 63 million years ago there was a sudden shift in reptilian history. Many of them died out. One plausible explanation offered is as follows. The reptiles are ectotherms *i.e.,* they get their body heat from direct exposure to sun's rays. They have no mechanism to generate heat internally like the mammals and birds do, which succeeded them in evolutionary history. A meteorite might have collided with the earth at that remote past raising a huge dust cloud that might have come in the way of sun's rays reaching the earth. Reptiles being ectotherms need sun light to heat their bodies. Excessive heat gives them a sun stroke. Similarly excessive cold leaves them inactive and inert. If this continues for a few days they lose their body heat and die. Scientists believe some such calamity might have sent the reptiles to their extinction. Only three types of reptiles survive today. The crocodiles, the lizards and the land living tortoises and the water living turtles. Some lizards lost their limbs gradually and became snakes. So the great reptilian line came to an end after dominating the earth for 167 million years during the Mesozoic Era between 230 to 63 million years ago.

We have seen that the reptiles were ectothermic and their inability to adjust to the fast changing climate proved to be their undoing. On the other hand the birds and the mammals which evolved from the reptiles and the evolutionary adaptations they made proved to be quite successful to become the dominant life in the air and on land during the last 63 million years. The birds took to the air to exploit the insect food available there leaving the land to the mammals. The birds and mammals are endotherms *i.e.,* they generate their heat internally. They keep their body temperature at constant levels usually between 36 degrees and 39 degrees centigrade. Constant body temperature makes it possible to evolve delicate and complex organs which greatly enhance the adaptive abilities of the birds and mammals. Birds with feathers and mammals with fur and a layer of fat under the skin could retain the body heat much better. Consequently they could function in the night or even in cold conditions where reptiles can not survive.

To maintain their bodies at constant temperature the mammals require a high rate of metabolism and a great deal of food to sustain it. Nearly 80% of the calories produced from our food goes to the maintenance of constant body temperature. This means increased in take of food as well as speedy extraction and metabolism of the nourishment. This was achieved by evolving special dentition. The reptilian teeth are peg like and slightly curved inward. They could only gulp their prey and swallow it whole. The digestion takes prolonged time. The mammalian teeth have three incisors, a canine, four premolars and three molars on each side in both upper and lower jaws expressed by the notation 3-1-4-3.

Even though this is the general rule there are some special cases. These specialized teeth cut, mash and grind the food mechanically before digestion. This helps to extract the nutrition from the food more efficiently and speedily.

But the most important innovation made by the mammals is neither the endothermic bodies nor the specialised dentition to sustain a high metabolic rate. They suckle the young with a mamma a breast which gave the mammals their name. They supply their young with an easily digestible food, the milk. They take care of their young as no other group did so far in the evolution. This allowed them to produce only a few young ones at a time. The fish, the amphibians and the reptiles laid a large number of eggs and left them. Many of them do not hatch, those which hatch do not survive their predators. The mammals on the other hand produced a few offspring and protected them. This ultimately led to their dominance during the Cenozoic Era which started 63 million years ago.

We have noted the main evolutionary innovations made by mammals. Let us have a brief survey of the conditions prevailed on the earth during this period which have made possible the birds and the mammals to evolve during this period.

There was not much change in the relative positions of the continents during the Cenozoic Period *i.e.,* during the last 65 million years. North America and Asia were connected at the Bering Straits. North and South American Continents were also connected to be separated latter and reconnected again. Eurasia and Africa were connected at the middle east. Australia remained an island continent during this period.

During the latter Mesozoic times and early Cenozoic times the continents were lifted up by the Earth's undulations. Earlier the continents were partially submerged by shallow seas. These crustal uplifts gave rise to modern mountain systems. The Alps, the Himalayas, the Rocky mountains in North America and the Andees in South America were formed during this period and to a certain extent the mountain building is still going on.

The uplift of continental masses and the rise of the mountains had developed new climate Zones. Changes in climate also effected the evolution of life. During most of the Mesozoic times the climate was tropical with uniform temperatures or with slight changes in temperature. As a result tropical plants extended to northern latitudes and Dinosaurs roamed the world freely. We have dinosaur fossils in almost all continents now. However as the mountain systems developed there was a gradual change into a variety of climatic zones. As time went on these have become regular seasonal variations and changes, with one season following the other in a cyclical fashion.

Such changes in climate was followed by changes in plant and animal life also. In the place of evergreen plants, deciduous plants made their appearance. They shed leaves regularly during a season. Flowering trees and grasses appeared which provided new habitats to insects, birds and animals. Many hoofed animals evolved to exploit these grasses and carnivores too to prey on them. All these developments resulted in the dominance of mammals during the last 65 million years as against the dominance of reptiles in the earlier 100 million years of Mesozoic Era. We will now follow the various stages of mammalian evolution. We have already noted the chief characteristics of mammals. Evolution is a continuous story. We make it compartmental to suit our interpretation.

There was no doubt of an abrupt decline in dinosaur fortunes some 65 million years ago. As already noted some scientists speculate that an asteroid might have smashed on to the earth and covered its atmosphere with dust for days on end making it difficult for the sun's rays to penetrate through the dust cloud. The ectothermic reptiles can not survive many days without the heating of sun's rays. So there was a mass extinction of the reptiles. Then the lowly mammals took over the empty spaces and radiated into the various ecological niches provided by the built up of mountain ranges climatic zones and variation in temperature, the evolution of various deciduous and flowering plants and grasses etc. It may not have been so dramatic but all of these factors contributed to the decline of reptiles and the multifarious evolution of mammals during the last 60 or 70 million years.

Mammals, however, did not appear from nowhere suddenly. They were there as long ago as 200 million years when the reptiles were at the height of their evolution. They were small shrew like animals probably nocturnal and insect eating. A small group of reptiles developed long slim leg bones. Some among them had a secondary palate which allows them to breath while eating. This is certainly not a reptilian feature. It had also developed the specialized teeth of a mammal. Among the descendants of this group were some small animals with uniform body temperature.

Differential dentition and endothermic bodies and a layer of fat under the skin, hair or fur are not the only characters of mammals which made them unique. It is the innovations they made in their reproductive system that made them unique. The early mammals experimented with three types of reproductive systems. Some laid eggs in the fashion of modern monotremes, the platypus or duck bill of Australia and the spiny ant eater of New Guinea. They also secrete milk from modified sweat glands under their belly which the young one lap up. These animals are not very successful however.

The second type of mammals are the marsupials so named because they possess a marsupium, a pouch outside their bellies in which they rear their young. The most popularly known examples are the Australian Kangaroo and the American opossum. The young in these animals are given birth to at a very early stage of growth after they were fertilized. Then they clamber on into their mother's pouch and attach themself to a teat nourishing on the milk of the mother. They remain there until they are sufficiently developed to forage on their own. Although there are several marsupial animals in Australia and some in South and North American Continents, their peculiar reproductive system restricted their success as compared to the placental mammals.

The placental mammals are the most dominant mammals now in almost all parts of the world. In the placental mammals the young develop inside the mother's womb called uterus. A thin membrane covering the developing embryo is in contact with the mother's uterus wall and through this area of contact, the placenta, food and Oxygen are carried from the mother to the developing embryo. The waste products of the embryo are also excreted through this mechanism via the mother's excretory organs. As a result of this prenatal growth of the placentals they are much more advanced at birth than the larva like new born of the marsupials. Hence in the competition between placentals and marsupials the latter were wiped out from the continental areas except Australia which remained an isolated continent during the last 70 million years.

As a result of all these developements warm bloodedness, specialized teeth for different types of food and above all care of the young with milk allowed the mammals to radiate into the nook and corners of this planet earth and adapt themselves to the various climatic zones. They have developed much better sense organs than the reptiles with which they could interact with their environments in a better manner. Consequently the brain matter also increased and intelligence developed in mammals. Man is a mammal and is heir to all these developments.

7

PRIMATE EVOLUTION

Monkeys which are called Primates in scientific terminology are a branch of mammals. Their order consists of 200 species including man. They are divided into three stages of development. The prosimians are the earliest group. They are the pre monkeys and include tree shrews, lorises , tarsiers and lemurs. Next came the monkeys, separated into the new world monkeys of South America and the old world monkeys of Asia and Africa. Then came the apes and finally man.

Apes are of four kinds — the Gorillas and Chimpanzees of Africa and the Gibbons and Orangutans of South East Asia. Man's early evolution took place largely in the savannas of Africa.

We humans have a great deal in common with the primates. The evolutionary developments made by this order primates directly benefited us. There are many anatomical and physiological similarities. Studies in biochemistry, cell structure, blood groups and genetics show close relationships between man and other primates. Not only physiologically but also in their social behaviour also there are many similarities between primates and man. Man developed and refined them much better. The primates live in groups which are often hierarchical, they communicate with each other through a variety of sounds and bodily gestures. The young ones have to learn all these behaviour patterns much like the young of humans. So a study of primate evolution throws much light on human evolution too.

Primates are descended from insectivore ancestors as long ago as 70 million years. We have fossils to confirm this. Why they have separated from insectivores depends on their tree living habits which prompted them to depend on the fruits and leaves and buds which are abundant in the tropical forests of the day. They have evolved to be masters of the trees for the abundant food there as well as the safety from their predators. These adaptations changed the anatomy and physiology of the primates in a remarkable way.

These changes include the development of forelimbs for grasping branches with an opposable first digit to the other four digits and nails on the fingers instead of claws. Along with this ability goes the ability to see where the animal is going with a stereoscopic vision. That means it should be able to judge the distance accurately, otherwise it suffers grievous falls. Stereoscopic vision is possible only when the eyes are in a frontal position unlike in the sides as in other mammals. The sense of smell is much reduced and consequently the snout is reduced. The brain area where the information from hand and eye are coordinated has increased. These developments increased the brain power of the primates. The prosimians or early monkeys give birth to more than one at a time. But their successors the monkeys and apes give birth to only one at a time and take care of them for a longer period.

These developments and refinements took place between 70 million to 15 million years ago. They made the primates masters of the trees. The geological upheavals produced tropical world ranging from Africa to middle east to Asia. Again some geological developments produced dry conditions and the forests began to retreat to their present positions some 35 million years ago. So some of the monkeys were forced to adapt a ground dwelling life at the edges of the forests. The great apes and finally man evolved from such ground dwelling species of primates. We will now take a detailed look at these developments.

The principal orders of mammals were already evident some 70 to 60 million years ago. They are the primates which exploited the way of life available in the emerging forests and the trees there. The rodents exploited life of burrowing, catching worms, seeds and nuts. The ungulates or hoofed animals exploited the leaves and more so the grasses and the carnivores preyed on the above three particularly the ungulates.

The evolution of primates is the evolution of the successive species with improved abilities to move about in trees and make a living there. The prosimians or half monkeys evolved from shrew like insectivores some 60 million years ago. They dominated the life in the trees for millions of years until the appearance of monkeys. They were present in the North and South American continents and Europe, Africa and Asia. Monkeys evolved some 34 million years ago and they ended the dominance of prosimians in Europe and North and South America. Only some nocturnal varieties survive in Asia and Africa. There they do not compete with monkeys which forage during day time. The island of Madagascar got separated from Africa and some species of lemurs survive there without competition from monkeys or any other predators.

Prosimians evolved some 70 to 60 million years ago from shrew like insectivore animals. They dominated the life in the trees for nearly 35 million years between 70-35 millions. Then between 35-25 million years monkeys and apes evolved from prosimians and their dominance in the trees was replaced by monkeys and apes. There are two evolutionary innovations made by prosimians to exploit the opportunities present in a tree dwelling life. The first one is a sure grip to their fore limbs by evolving an opposable thumb to the other four fingers. The fingers also had nails instead of claws. They grasp the branches with their four fingers on one side and the thumb on the other side in a power grip as when we handle a hammer. This gives them a sure grip in the branches of the trees.

The second innovation was vision. Mammals generally have their eyes on either side of the muzzle. This gives them wider field of vision but not a sure depth perception. For that the two eyes have to move to a single plane. The prosimians achieved this by shifting their eyes to the front of their face instead of the sides. This produced a stereoscopic vision which allowed depth perception. This allowed the prosimians to estimate how far the next branch was. Without such perception it was not possible for them to move in the trees jumping from one branch to another. Thus for the first time in mammals the sense of vision gained prominence over the sense of smell which was prominent in all other orders of mammals. This resulted in the reduction of the size of the snout and also allowed for a greater volume of the brain. Also the area devoted to analyzing the smells in the brain is reduced and the area devoted to analyzing the visual information increased. This also increased the area in the brain which is devoted to the coordination of visual information with muscular activity. The complexity in the nervous circuits in the brain as well as the brain size increased.

By 35 million years ago monkeys began to evolve from prosimians. There were two main branches. One is the American branch and the other is African and Asian branch. These went their separate ways in latter evolution. They inherited the developments made by the prosimians and improved upon them. The prosimians were mostly nocturnal. They had no colour vision. The monkeys were day time living. Colour vision was of great advantage to them. So they developed colour vision. Their hands also developed individual manipulation of fingers. This is known as precision grip as when we handle a pen or pencil. Monkeys regularly groom each other and could handle strange objects with their hands and examine them with their eyes. This gives them more accurate information about their environment as against sniffing with their noses which is the principal information gathering sense organ for other mammals. The monkey's way of vision, smell, taste, touch and hearing have not changed during the last 35 million years. They have gathered a vast repertoire of information and to store and integrate all that information they needed a large brain capacity. So the monkeys advanced over the prosimians and remained till date the dominant species in the trees.

The apes evolved some 25-20 million years ago, either from monkeys or prosimians is not clear. In the African jungles south of the Sahara live Gorillas and Chimpanjees in groups ranging from 10 to 50 to a group. In the forests of south east Asia live the Gibbons and in the forests of Borneo and Sumatra live the Orang-utan. All the apes are large in size except the Gibbon. They have longer hands than legs. Their locomotion is largely by brachiation, *i.e.,* swinging from tree branches with their large and powerful hands.

The Gorillas and Chimpanzees could also walk on all fours on ground, but while moving about they lean on their knuckles of the fore arms. They could also stand on two legs and walk short distances. The Gibbons also could walk short distances balancing with their hands on their head. The Orang-utans are solitary creatures which rarely come down to the ground. They are heavy in size and they distribute their weight on three or four branches using both hands and legs for the purpose. For this purpose the apes and monkeys also developed an opposable great toe to the other four fingers in their feet. Thus they could manipulate all their limbs to move about in trees.

The apes and monkeys differ anatomically a great deal. Monkeys move on all fours either on tree trunks or on land. That gave them a different shape of body and internal arrangement of viscera. The apes in their brachiating habits developed a different anatomy particularly between the pelvis and head. This shape is almost human like. They could swing their hands much more freely and in a round about position. Whereas monkeys move their hands or feet only in a backward and forward position. For an upright position man developed a wider pelvic girdle and the complicated muscular structure to hold the torso. This the apes lack and consequently they cannot stand for long. Finally between 9 and 6 million years ago first the branch of gorillas and next the branch that led to Chimpanzees separated from the branch that led to humans.

We have seen in the previous chapters that at each stage of the animal evolution they have acquired some special features and they have passed on those successful characteristics to the succeeding forms of evolution. We will examine briefly what are the special characteristics evolved by primates and passed on to man.

The primate skull sits vertically on the vertebral column and the nervous chord passes through the hole known as the foramen magnum into the skull in a central position. This central position of foramen magnum at the base of the skull is typical of all animals who could sit or stand upright. This gives the animal a straight face where the brain case is large and above the face. The sense of smell is replaced by the sense of vision and consequently the size of the muzzle is much reduced and the eyes are shifted to a frontal plane which provided both stereoscopic and colour vision. For stereoscopic vision two images of objects slightly overlapping each other are required. This is provided by the two eyes set in the same frontal plane apart from each other. The images produced by them overlap in the brain there by producing a three dimensional effect. Another important feature of the primate vision is they possess colour vision also. This is made possible by the evolution of rods and cones at the back of the retina. Rods end in flat tips and are sensitive only to monochromatic light. Most animals including the prosimians possess only monochromatic vision. It is the monkeys which have evolved colour vision. This is made possible by what are known as cones at the back of the retina that end in sharp tips. They are sensitive to various wave lengths of polychromatic light. They also have a special feature in the retina of each eye popularly known as the yellow spot. The cones are also responsible for this. This allows the animal to focus attention on an object without losing touch of its surroundings as when you are reading this page you could focus on a particular word without losing the rest of the page from your vision.

Primate hands have a power grip with an opposable thumb to the other fingers as well as a pencil grip where they could hold small objects with thumb and forefinger and bring them close to the eyes for a very detailed examination. The fingers have nails and have sensitive pads underneath . They enhance their sense of touch. The feet also with various degrees of modifications allowed them to stand upright even for short periods.

Primate brain evolved concurrently along with the evolution of its hands for grasping and its vision for stereoscopic and colour vision. In the lower orders of mammals the development of brain is limited to brain stem and midbrain which coordinates the muscles and maintains the basic bodily functions like breathing, blood circulation, metabolism and maintenance of body temperature etc., without which the mammals can not survive. With the primates the function of the brain is much enlarged and so the simple brains of the other orders of mammals is not sufficient to cope up with the much enhanced and elaborate activities of the primates. Living in the trees demands a constant interaction with its environment as well as a fast response to its sense receptors, particularly vision and a quick response to its locomotor limbs. Otherwise primates can not move through the branches of the trees with all their acrobatics without suffering serious falls. So integration of the information gathered from its sense organs and sending proper orders to its locomotor limbs is at the root of primate's success in their arboreal life.

In this process vision played a principal role above all other sense organs like smell, hearing, touch etc. In the lower orders of mammals it is the sense of smell which played primary role in its senses. That means they have to be constantly sniffing at objects lowering their head. In primates they could hold their head high and look around them and thereby obtaining information from a wider area and thus command a lot more information than other mammals. More information means more brain matter to store it and integrate it in a proper manner. Not only that, the primates could hold even small objects with their thumb and forefinger and bring it before their eyes for detailed observation thus obtaining more information about its surroundings than any other mammal.

Thus the monkey's and ape's way of interaction and reaction to their surroundings is a completely new development in the evolutionary history of mammals and this innovation was passed on to us humans who shared ancestors with gorillas and chimpanzees some 9 and 6 million years ago respectively. It is the brain power which set humans apart from other animals and the roots to such development lay in the evolution of monkeys and apes and their arboreal living for nearly 60 or 70 million years, which had brought about radically different changes in the anatomy and physiology and the way they interacted with their environment. It is ultimately an organism's ability to interact with its environment to obtain food, protect from predators and live long enough to produce offspring that determines the success of that organism. In this respect the order of primates particularly, the monkeys and apes are quite successfully adapted and we humans are better than them only in degree. Alas that proved to be their undoing. We are destroying their habitats indiscriminately.

Behaviour *i.e.*, the adjustment to the other members of the group or the environment is both instinctive as well as learned. The nearer the organism is to the lower levels of evolution the more the instinctive part and less the learned part. Conversely the nearer it is to humans, the less is the instinctive part and more is the learned part. Learning is generally imparted by the mother to the child. This necessarily involves a strong mother child bond. The child of a monkey depends on its mother for one or two years. The child of a chimpanzee or gorilla depends on its mother for 3 to 4 years. Whereas the dependence of a human child on its mother is the longest 7 to 8 years.

We have seen in the lower orders of organisms the mother child bond is minimal or non-existent. Fishes and oysters lay thousands of eggs and just leave them. Even amphibians do the same though they are not so prolific. The reptiles lay fewer eggs and their care of the young is minimal. It is only in mammals they lay litters in some species and only one or two in the larger animals. It is in the higher primates we see that the mother gives birth to one at a time. The dependence of the child on the mother is also longest as the order progresses towards humans. The mother child bond is also progressively stronger and more emotional without which it would well neigh be impossible for the mother to take care of its young against many odds and the young one to survive. The long period of dependence helps the child to learn behaviour suitable to adjust itself among its group as well as in its environment. The learned part of the behaviour increases and the instinctive part decreases. This is how the primates come to adjust themselves in their group and fill the role appropriate for them. They learn this behaviour first from their mothers, latter from their siblings and much latter from their play mates and peers and finally from the other members of the group. They learn the hierarchical nature of the group and fill the place expected of them and whenever an opportunity presents as they grow up they also try to move up the hierarchy of dominance. The dominant male is usually the leader of the group.

Primatology, the study of primate behaviour has become a very important field of study during the second half of 20th Century. Extensive studies were made both in the laboratory as well as in the natural field conditions. The primates live in groups. The number in the group varying from species to species. The group is the custodian of primate behaviour and it is passed on to the next generation. The infant born into the group receives the protection of the group as well as its learning for which there is ample time through its slow maturation. Apart from the protection of the mother during infancy, it is the play with other siblings and playmates that teaches a young monkey or ape the proper behaviour with others of the group. In laboratory experiments the monkeys which did not have play were unable to adjust themselves to other monkeys. Obviously it is the play with other growing up monkeys that produced a well adjusted monkey in its social group. When they become adults both females and males are equipped to fill their roles in their group. Thus the primates receive their schooling in their group living and they learn to fill their appropriate roles in the group and gradually replace the older females and males in the hierarchical order of their group.

All animals of a species can communicate with varying degrees of articulation. The primates also can do so to a much better extent. They have a number of calls which could express together with facial and bodily movements a variety of messages. They include, among others, warning calls, threat calls, defence calls and gathering calls. Grooming each other is a particular pastime of primates and indicate submissiveness and friendliness.

Several attempts were made to teach chimpanzees and gorillas human language. The attempts failed because they lacked in their brains the connections necessary between their auditory areas and vocal areas. Their vocal cords in the throat, mouth, tongue and lips also lacked the flexibility to produce the variety of sounds that a human could produce.

Better results were produced when the American sign language for the deaf was taught to them. The Chimpanzee named Washoe and the Gorilla named Koko learned several words through this method and they were even able to produce short sentences using such sign language.

We have learned the possibilities as well as the limitations of primates to becoming human. They are clearly the stepping stones on which we humans have evolved. The primary sense organs of vision, smell, hearing taste and touch have evolved in monkeys and apes 25 million years ago and humans inherited them and there was not much improvement. The improvement took place in humans in content and volume and its storage and assimilation of the infromation gathered by these sense organs. That requires an enormous increment in the cortex which is the storage and thinking part of the brain. Chimpanzee's brain capacity is less than 400 c.c., whereas human brain capacity is upwards of 1300 c.c.

Other areas where humans differed markedly are in their ability to stand upright and walk on two feet. This involves the widening of his pelvic girdle and the strengthening and altering of his muscle structure which supports the torso. This also involves the change in the arrangement of internal viscera particularly in the urogenitary and reproductive organs. Bipedal locomotion allowed the hands to be free and man gradually learned to make and use tools with them. This required a level of conceptual thought which is not to be found in monkeys and apes.

Conceptual thought gives man language and the power of cumulation to his experiences. Monkeys and apes could transmit to their young the same kind of behaviour generation after generation. In humans this behaviour is moulded on the accumulation of knowledge from one generation to the next. This gives humans culture, civilization, art and architecture, music dance and drama. Religion develops, morals, ethics and law evolve together with mathematics, logic science and technology. Man has also become heir to passions which are base as well as noble. All these developments stood man apart from monkeys and apes not only in degree but in kind. Man is distinctly separate from monkeys and apes though he shared ancestors with them during the last 25 million to 5 million years. Since 5 million years ago the stem that developed into mankind had separated and evolved to man. We will study the developments in the succeeding chapters of this book.

8

HUMAN EVOLUTION: AUSTRALOPITHECUS

In the Miocene period 25-15 million years ago there were broad stretches of continuous forests from Africa through Europe, Asia towards South East Asia. In these forests many species of monkeys and apes evolved and proliferated. Towards the end of this period and the beginning of the next period Pliocene 15-2 million, the atmospheric conditions changed to cool and dry weather and the forest cover dwindled and shrank towards the equator. So some of the apes and monkeys were forced to come down on to the savanna at the edge of the forests. Some such groups the macaques and baboons of Africa, the rhesus monkeys of India, the Chimpanzees and Gorillas of Africa still follow such a life style. They forage on the ground and when threatened climb on to the trees.

A fragment of a jaw was found by G.E. Lewis an American Anthropologist in the Siwalik Hills of Punjab in the 1930s. He named it Ramapithecus Punjabicus and claimed it to be on an ancestral line of one of those hominids which eventually led to man. This is a bold claim based only on a fragment of a jaw. But often Anthropologists have to work with some such fragments of fossils only.

Lewis' claim rests on two important clues. The roof of the palate that is the upper portion of the jaw is curved like an arch. Secondly the row of teeth widened towards the back of the molars in a parabola shape. Thirdly the canine is small and the premolars even. In the apes the canines are large the rows of teeth on both sides are parallel forming a U shape and the roof of the palate is flat. These differences indicate that Ramapithecus was a hominid or on the way of becoming one. But we do not know whether it walked erect or not since no leg, hip or spinal bones are available.

L.S.B Leakey found another piece of the upper jaw in Africa and what is more the Indian and the African jaws matched, indicating that the range of Ramapithecus was very wide. The age of the African Jaw was determined by potassium argon method and the age given was 14 million years. There was no continuity of hominid fossil record. The long period of Pliocene 14 to 2 million is a blank as far as Hominid fossil record is concerned. Using other techniques like genetical comparison, blood groups, blood proteins and molecular biology techniques, scientists estimate the ancestors of modern gorillas and chimpanzees and of man branched out some 6 to 9 million years ago. Then how Ramapithecus could exhibit hominid features 14 million years ago is a puzzle. Unless more hominid fossils dating between 14 million and 3 million are found, this puzzle can not be solved. Until then Ramaphithecus has to remain as an odd man out.

In 1925 Raymond Dart of South Africa came up with a skull of a baby hominid. He named it Australopithecus or simply the South Africa Ape. Though it was the skull of a baby hominid its brain capacity was larger than that of chimpanzees. It had milk teeth only. Yet they were more like a human child's teeth. The incisors and canines were smaller compared to the apes. They were larger in apes and monkeys. Its foramen magnum through which the nerves pass into the brain is forward and downward under the skull cap. That indicates the Australopithecus baby was walking erect. In apes and monkeys the foramen magnum is more towards the back indicating a forward hanging head and a horizontal spine. However Dart did not have any other bones more particularly those of leg and pelvic bones which are necessary to prove the erect posture of Taung baby so named after the place it was found.

Robert Broom set out to find more evidence to support Dart's Taung baby. They lacked any adult skull or leg and pelvic bones. Broom soon could get several fragments of skulls from two different sites as well as many leg and pelvic bones. These bones when arranged proved to be from two different species. The second species Broom named Paranthropus, the first species was of course Australopithecus. Broom and his colleague Robinson found several fragments of these two species as well as that of an advanced Australopithecus species. The Paranthropus was robust with massive jaws and skull ridges to support the muscles that worked the massive jaws. It is erect in posture. It remained a vegetarian. It did not make any tools. Ultimately its line became extinct. The Australopithecus also walked erect. It evolved further to make tools. It was not as robust in build as Paranthropus. Its latter forms made and learned to use tools by about 2 million years. These finds were also supported by L.S.B. Leakey and his wife Mary's finds in East Africa in Olduvai Gorge.

Olduvai George is a gash in the Earth some 25 miles long and 300 feet deep in the northern Tanzania cut by some ancient river. It was first discovered by a German naturalist Hans Reck in 1913. Its remoteness and inaccessibility prevented its study by early anthropologists. L.S.B. Leakey was the son of an English missionary in East Africa. He was born and brought up in Africa knowing the cultures, the people and the language of the natives well. He along with his wife Mary decided to explore this gorge in the early 1930s. The gorge itself contained four distinct beds or layers containing different species of fauna in each bed. The dating followed by geologists so far of such beds was by stratigraphic methods. But African stratigraphy was not well-known. The fauna belonged to one period in Europe belonged to another period in Africa. The differences in temperature and humidity and rain produced these differences. Accurate dating had to wait until after World War II when radio active methods were developed. Even then the fossils had to be embedded in materials containing radio active substances. Fortunately for Olduvai Gorge, it had many layers of volcanic ash and their dates could be accurately determined. They ranged from 2 million to 1/2 million.

Leakey's first expedition produced specimens of fossils from some 150 animals, many of them are extinct. They came from all the four beds. They conducted a second expedition in 1959 with more finances available. Even in the first expedition in the 1930s, many stone tools were found in the lowest bed. They were round pebbles of many sizes but one end chipped away to make a jagged edge and the rounded side could be fitted into the palm. But no fossils were found who made these tools. In his second expedition he was able to find fossils of a jaw and part of a skull most of the face and all teeth. Tools were also found with it. It was found between two layers of volcanic ash and dating it was made easy. It was dated at 2 million years. This skull was given the name Zinjanthropus. It was similar to the Paranthropus fossils found in South Africa.

In their subsequent expeditions, Leakeys found another skull which was more like that of Australopithecus. It was also associated with tools and in the same bottom bed giving it the age of 2 million years. It was named Homo habilis latter. Several specimens of fossil ape men were dug out subsequently which were dated between 2 million and 500 thousand years. While there was no change in the Paranthropus there was marked evolutionary change in Australopithecus. The advanced models approximated to Homo Erectus status which we will discuss in the next chapter.

Often when there is an innovation in evolution it is taken advantage of and several species are formed. Similarly when the Miocene apes evolved erect posture somewhere during the Pliocene during 14 to 3 million years, there evolved several erect and semi erect apes. Some of them further evolved and finally led onwards towards man. Others which did not evolve further either became extinct or remained as another species of apes like the present day Chimpanzees and Gorillas. Paranthropus is one such extinct species. It lived on between 2 million and 500 thousand years without evolution as its habitat remained intact and Paranthropus found little spur for evolution. It happily chewed on the berries and roots of the forests as long as the going was good. When faced with competition and altered conditions it could not survive. So we are left with Australopithecus dominating the scene. Leakey suggests that it was Homo habilis which further evolved towards man and the tools found at Olduvai Gorge aged 2 million years were made by this Homo habilis and not Zinjanthropus.

In the Hadar region of Ethiopia in 1974, Donald C. Johanson found the partial skeleton of a female. It lacked the head, the most vital part of any species of fossils. It was named Lucy. It had long arms like an ape indicating brachiating ability. Its pelvic and leg bones indicate that it walked on two legs. Her age was estimated at 3.18 million years. It is believed to be the common ancestor of latter hominids. In 1992 Johanson's team organized another expedition to the Hadar region of Ethiopia to find out the skull which Lucy's remains lacked. They found the skull of a male with "a jutting jaw, heavy brow, flaring cheeks, and a sloping forehead". The skull capacity was estimated at more than 500 c.c. The species was known as Australopithecus Afarensis.

There are however marked differences between Afarensis and the South African specimens of Australopithecus Africaners found by Dart, Broom and Robinson. Similarly there are differences with Zinjanthropus and Homo habilis specimens found in Olduvai Gorge by the Leakey family.

Obviously the conclusion is that after 3 million years there was a marked change in the climate of Africa. As a result the Afarensis or whatever is the ancestral type gave rise to several branches of new hominids. One of them evolved further into Homo and the others like Paranthropus, Zinjanthropus etc., became extinct. Lee Berger writes (National Geographic, August 1998) that scientists have identified at least seven species of hominids that evolved between 4 million and 2.5 million and one of which led to our genus Homo and the others were extinct.

So we shall see what the ancestors of man were like between 3 million and 1 million years ago. The first candidate was of course Lucy found in the Hadar region of Ethiopia by Donald C. Johanson. By his account it walked erect. Erect walking requires changes in spinal bones, hip and pelvic bones, leg bones and foot bones. Each half of the pelvic girdle consists of three bones, the ilium the wide hip bone, the ischium the lower part of the hip which takes the weight when we sit down and the pubis which is the joining bone, in the front with the other half. The three bones join in a cup in adults and this is called the acetabulum. The head of the thigh bone known as femur fits into this cup and allows for the free rotation of the leg. The pelvic bones attaches to the powerful muscles needed for upright gait, to the spine on the top and to the thigh bone below. So for the determination that a hominid walked upright or not an examination of the bones of the pelvis is essential. Fortunately Lucy's pelvic as well as leg bones were discovered. So we could deduce that it walked upright.

Other important bones to deduce bipedal locomotion are the leg bones and foot bones. The long bone of the upper arm is known as humerus and the long and strong bone of the thigh is known as femur. The ends of these bones which joins up with the body indicate the mode of locomotion of the animal whether it is a quadrupedal or bipedal. We have seen that the ball shaped end of the thigh bone fits into the cup shaped acetabulum of the pelvic girdle.

The foot bones also indicate whether they belong to the bipedal man or ape or monkey. It has a lot to do with locomotion. The foot of bipedal man has to support the weight which is mostly done by the heel, the arch and the big toe. In apes and monkeys the big toe is not in line with the other toes but away from them and opposable to them to various degrees. It is used in climbing trees and grasping branches. It is in line with the other toes in hominids and the metacarpal bone behind the big toe is in line with it and quite strong. Leakey found such fossil of a foot at Olduvai Gorge in Tanzania and its age was estimated at 2 million years.

Though there were practically no fossil vertebral columns present, we should mention the differences between the vertebral columns of a quadrupedal and a bipedal. In the quadrupedals it is bent like the arch of a bow and in the bipedals it is bent like an S curve. The lumbar region *i.e.*, at the belly it is bent forward, at the thoracic region *i.e.*, at the lungs it is bent backwards and at the cervical region *i.e.*, at the neck it is straight.

The skull is also a very important source to determine whether its owner walked on all fours or a bipedal. The nervous chord goes, via the vertebral column and enters the brain through a hole at the base of the skull. This hole is known as the foramen magnum and it occupies a central position in man and other hominids who walked erect on two feet. It moves backward as the animal is a brachiator like the apes or a semi quadruped as the monkeys are or fully quadruped like the prosimians are. The backward movement of the foramen magnum in the base of the skull is an indicator of the distance between the humans and other primates in the evolutionary scale. So we have several indicators in the bones of the species to show whether they walked upright or not. By these indicators we know that Lucy of Ethiopia which dates prior to 3 million years was a bipedal. There are however no stone tools that could be associated with the fossils of Lucy or its kind.

Meave Leakey, wife of Richard Leakey and daughter in law of L.S.B and Mary reports (National Gegraphic September, 1995) that she had found some significant fossil finds of a species of hominid at a place known as Kanopoi in the Lake Turkana region of Kenya. The fossils found were a part of the upper jaw and a full lower jaw and the lower and upper ends of the big leg bone between knee and ankle known as tibia. The leg bones indicate it was an erect walking creature. Its upper jaw or maxilla shows that it is a bit more ape like. Its age was estimated at 4.1 million years.

Tim White unearthed a nearly complete skeleton in Ethiopia at a site called Aramis which is dated at 4.4 million years.

Meave Leakey herself announced (National Geographic October, 2001) the discovery of a hominid skull from the Lake Turkana region of Kenya dated between 3.5 to 3.2 million years. Thus, paleo-anthropologists are excavating more and more fossil ancestors of man and extending our knowledge of our past history.

So the implication is that bipedal gait evolved earlier than tool using and tool making ability. However some anthropologists do not agree. They point out that stone tools are durable and hence we are able to find them. Tools made of perishable materials like wood or unshaped stones are also tools and perhaps these early hominids used such tools. What is more Jane Goodal* observed chimpanzees for many years in the Gombe reserve forest in Tanzania. She points out that chimpanzees use a long blade of grass or a vine

* Jane Van Lawick Goodal : In the Shadow of Man.

plucked of its leaves to extract termites from their mounds and then eat them. Not only that, the young chimpanzees learn the technique and behaviour from their mothers.

So while chimpanzees could use tools made of perishable materials why not the evolving hominids who by all means had larger brains. Then we should agree that one species of the Galapagos finch which uses cactus needles to extract insect grubs out of holes in tree trunks is also using tools, albeit of perishable materials.

So it is obvious that we have to draw a line somewhere. Now it is generally agreed that tools made to a design by a conscious and wilful effort imparted by some degree of conceptualization could only be treated as proper tools and such hominids are tool making and tool using hominids. We find such tools at the lowest bed of Olduvai Gorge discovered by L.S.B. Leakey in his first expedition to that place in the 1930s. Those tools are now called to belong to the Oldowan industry. They were dated at 2 million years and are believed to be made by Homo Habilis variety of Hominid.

The Oldowan tools are generally water worn pebbles which were struck with another pebble to remove some flakes from one end to create a jagged and sharp edge. The smooth end is fitted into the palm. The flakes also serve as sharp edged tools. These are quite effective. L.S.B. Leakey is reported to have skinned and dismembered a small antelope with such tools in 20 minutes. The next development in tool making is flake tools and bifaces. They are natural extensions of Oldowan tools. The flakes removed from one end of the pebble is flaked all round, then it becomes a biface. It could be a good pounding instrument. Between 2 million and 1 million years there is not much development in the tools than the above. The raw material quartz and quartzite which is abundantly available in Africa may also have acted as a limiting factor. Flint, which is available in Europe is required to make tools with much sharper edges.

The tool making ability though looks simple to us now, brought about profound changes in the life of evolving hominids. The changes in their life which were forced on them by changing environment gave them a selection pressure to make and use tools.

Shaped stone tools are not necessarily the first tools used by them. They could have used sticks or leg bones of other animals particularly antelopes to club their prey or predators. They could have used unshaped stones to break open bones for the marrow inside. Carnivores usually leave the head and legs of their prey as they are difficult to chew. The emerging hominids might have taken them away and have broken the bones open for the marrow and brains inside. It is a very nutritious food and would be a great help for the increasing brain matter of the hominids. Unshaped stones might have suggested that it would be helpful if one edge is jagged. Such implements also help them to kill and skin and butcher their prey since they can not do it with claws or canines which they lacked. What they lacked was made up by tools. Tool making though replaced claws and canines it helped emerge the brain and the nervous coordination between the sensory organs of vision and hearing and touch and the locomotor organs of hands fingers and legs and feet. Tool making required a brain which could conceptualize, which could symbolize, visualize the end product in its mental image and perform certain operations on it in the imagination. This is the uniqueness of the hominids which have achieved this capacity by 2 million years when its brain capacity was about 750 c.c. and perfected this capacity by about 1 million years by which time the brain capacity developed to 900 c.c.

So the evolution of the brain of the hominids between 2 million and 1 million had many other consequences. They are conceptualization and symbolization of external objects and events, hunting small animals on a regular basis from being a retriever of bovid remains from Carnivore meals probably in competition with vulture and hyena like animals. Living in small groups developed bonds between male female child stronger than in apes. Communication between individuals based on some sounds and gestures is essential for any cooperative group living and hunting. Lastly to depend on a cultural evolution rather than on physical evolution, which is introducing a unique feature into hitherto physical evolution, made Homos unique. The surviving lone species of this genus homos is ourselves Homo Sapiens and we are unique because our ancestors the Homos hit upon a trail blazing new path in the evolutionary history of living organisms. We will study them one by one.

Tool making requires at least a rudimentary capability of conceptualization. To identify similarities between various objects is the first step in conceptualization. Water worn stones are usually found in streams or on the edges of lakes. They are roughly of the same shape though of different sizes. These stones which are handy for throwing as well as useful to crack open bones and nuts must surely have been used earlier than seeing a tool in them by chipping at one end and using it with more effect. This is perhaps the beginning of conceptualization. This requires development of the fore brain as well as the associational areas of mid brain. The fore brain is the base of cortex which is responsible for memory, image formation, its retrieval and performing some mental operations on it or thought. Gorillas and Chimpanzees can not do this. For them what is not present to their vision does not exist. They can not retain any image for more than a few moments. That is the difference between hominids and apes as far as their brain development is concerned. We have already noted that erect posture evolved earlier than tool use and brain development in hominids. Tool use and brain development evolved between 2 million and 1 million years and the brain size also increased from some 700 cc to 900 cc. This development has other important consequences.

Tool use permitted the change in mostly vegetarian food habits of early hominids to regular meat eaters. Earlier they supplemented their vegetarian food with lizards, birds, small antelopes, rodents and the left overs of bovid heads and legs from carnivore kills. That was what is evident from the fossil remains of Bed I at Olduvai or the South African caves excavated by Broom and Robinson. The fossils of Bed II at Olduvai and excavations of similar age elsewhere showed fossil remains of larger antelopes and other mammals. They must surely have been hunted down and for hunting such large animals require cooperation, division of labour, bonding between males and females on a more permanent basis than was found in apes and lastly communication on a more extensive basis than in apes.

In apes and monkeys the cooperation is limited and the number of individuals in a group often vary. The cooperation is never extended to food sharing. In the evolving hominids the cooperation is extended to sharing food and division of labour. Often the males are the hunters and the females are the gatherers. Also the females with their children cannot take part in the laborious hunts. So food must be taken 'home' to a base camp, usually a rock over hang, or cave or simply a temporary shelter in the African savanna by the edge of a lake or river or stream. Several stone tools and the rock fragments were found at such places indicate that the group may consist roughly of 10 members like the present day inhabitants of the Kalahari desert.

The often permanent bonding between the male and female had its roots in the continuous sexuality of monkeys, apes and hominids. The females of other animals receive the males only at the oestrous. The monkey, ape, hominid primates had a modified oestrous cycle of 25 to 29 days period. This makes them continuously receptive to sexual activity and there is more tightly knit bonding between males and females. Added to that the hominid baby is more helpless at birth and requires a prolonged nursing and care taking by its mother. Above all the division of labour between the males and females between hunting and gathering activities and sharing of food and base camp have all set the hominids on a different course of evolution than the other primate groups.

The communication required for the above activities is not the simple sounds and gestures we find among the monkeys and apes. It is much more elaborate definitely. Unfortunately fossils do not indicate whether their owners had speech and language abilities. However developments in the skull in two areas indicate the development of speech capabilities. These are the Broca's area and the Wernicke's area. However we can not say confidently that the hominids between 2 million and up to 500 thousand years had any capacity for true speech.

All these changes were brought about by a slow shift from physical evolution to cultural evolution each acting on the other and reinforcing the other. Ultimately in modern man it is the culture which determines his living patterns and adjustments to his environment than the slow and often wasteful physical evolution which is all that is available to lower forms of living organisms. This set man apart from other organisms not only in degree but also in kind.

9
HOMO ERECTUS

When Darwin published his books 'Origin of Species' and 'Descent of Man', they raised a controversy as Darwin suggested that man as well as all other forms of life evolved from earlier forms of life by gradual modifications. Then there have to be some missing links between the humans and the great apes which closely resemble humans in the living world. Then who are they. Clearly they have to be extinct fossils. Then where could we find them. Darwin himself suggested that man's ancestors, the missing links between the apes and man could be found in Africa since the Gorillas and Chimpanzees who are close to him among the animal world are living in Africa. By the same analogy the missing links could as well be found in South East Asia where the other two great apes the orang-utans and the gibbons are living.

A Dutch anatomist who reasoned it so and joined the Dutch Army as army doctor went to Dutch East Indies determined to find out the missing links. His name was Eugene Dubois and the year was 1887. He started digging in the island of Java near a small village Trinil on the banks of the river Solo at a bend in the river. He used conscript prisoners for his labourers, since in those days scientific labours did not receive funds so generously as now. The first finds were fossils of animal bones which are extinct now. He finally came up with a tooth which looked like that of a human tooth. This was followed by a skull cap and next by a femur and another tooth. These finds were separated only by a few yards and Dubois concluded they came from the same individual. However the age of these fossils became a controversy as little was known about the geology of Java at that time. Dubois however came to the conclusion that these fossils represented a transitional form between man and ape and named it Pithecanthropus Erectus or Ape Man with erect posture.

The femur, the long thigh bone was quite erect and indistinguishable from human thigh bone. The pithecanthropus walked erect was not disputed. The trouble came with the skull. Its capacity was estimated at 850c.c. and the areas associated with speech were also present in a rudimentary form. The associational areas also did not develop to a human level. The sense centers were well developed.

The trouble with pithecanthropus was that it was almost like a human from neck down but not quite human in its skull. It was neither a human nor an ape. It was intermediary between the two. The fossils were too early for public as well as scientific opinion to accept them as they are and controversy raged for many years. Dubois, in disgust locked the bones and refused to show them to any one. He did not relent until 1924 when other scientists persuaded him to show his fossils for scientific study. He presented his original fossils as well as other specimens he discovered. His work was continued in Java by Von Koenigswald and he discovered some more fossils at places like Sangiran and Mojokerto which reinforced the findings of Dubois. The Java man was essentially an erect walking homo with thick brow ridges above the eyes, the forehead was receding and the skull cap flat. The volume of the brain was estimated at 900c.c.

During the occupation of the Dutch East Indies by Japanese army in 1941 Von Koenigswald had to secrete his fossil specimens with his friends and replace the originals with casts. After the war when he was released he was able to recover many of his specimens intact. However a different fate awaited the fossils of Sinanthropus Pekinensis, 'Peking man'. They were lost during the war. It was also an erect walking ape-man or Homo Erectus.

Sinanthropus or Peking man was a close relative of Pithecanthropus or Java man. Both exhibit similar features. In the year 1924 Dr. J. Gunnar Anderson and Dr. Zdansky who were working for Chinese Geological Survey found two human type molars investigating the limestone caves near Choukoutien (now Zhoukoudian) about 30 miles south of Peking (now Beijing). They handed over the molars to Dr. Davidson Black, professor of Anatomy at Union Medical College Peking, for study. He realized the importance of the molars and properly assigned them to a new genus named Sinanthropus Pekinensis or Peking man for short. The Rockfeller Foundation of New York agreed to finance the excavation of the caves and a large expedition was organized in 1927 for a scientific digging and study of the site. The work at site was supervised by professor C. Li assisted by Mr. Pei both of China. Several international scientists joined them from time to time. The overall study of the fossils was done by Davidson Black. While work was going on Davidson Black died in 1934 of a heart attack. He was replaced at the Union Medical College by Franz Weidenreich who studied the fossils latter.

When America declared war on Japan after Pearl Harbour attack in 1941 the fossils were packed hurriedly for shipment to U.S.A. They were lost in the melee whether on land or at sea is not known. However Weidenreich had sent casts and detailed descriptions and figures to U.S.A. in advance and they survived. Weidenreich being a German national was interned in U.S.A. He spent the rest of his life studying and publishing the material he sent to U.S.A. After the war when the communists took over the Government in China they accused the United States of secreting part of their national treasure *i.e.*, the fossils of Peking man, in one of their Natural History Museums. The Americans hotly denied the charge. There rests the matter of the fossils of Peking man.

The stories of Java man and Peking man are important milestones in the study of the prehistory of man. The stories contributed by Dart, Broom and Robinson about the South African hominids and the contributions of the Leakey family in East Africa are other milestones in man's prehistory. They evoked tremendous interest in Paleoanthropology. After the war several expeditions were organised by many Universities and Natural History Museums with adequate funding. Now any excavation is a multi-disciplinary activity and they have shed much light on man's prehistory. Due to their efforts, we now know that Homo Erectus lived not only in Java and China but in Africa itself and Europe also. With its large brain and near perfect erectness similar to modern man it has acquired much better organizational and communicative skills as well as better tools and extended its range out of Africa and spread to tropical as well as sub-tropical regions.

Rick Gore* describes for us those postwar developments with those inimitable photographs of National Geographic by Kenneth Garrett. Italian archeologists collected several fragments of a skull cap at a place called Ceprano south east of Rome at a highway construction site. They reassembled the skull cap by 1996. It has "low sloping forehead and large browridges" the hallmarks of Homo Erectus. Its age was estimated at 8,00,000 to 9,00,000 years.

At about the same time British scientists discovered the lower leg bone, tibia at a place called Boxgrove a few miles away from the English Channel. The bone was quite massive, double as thick as that of a modern human's and estimated to be of 5,00,000 years old.

At about the same time in 1994 at a place called Gran Dolino in Spain archeologists have dug up about a hundred fragments of what are to be Homo Erectus fossil fragments and about 200 stone tools and other animal bones. They were estimated to be aged about 8,00,000 years old.

There are other finds of wooden spears and bone tools in Germany and finds of graves in Spain. But their dates indicate they might be transitional forms between Homo Erectus and Neandertals which we will discuss in the next chapter. Even the ages given to the fossil finds in Italy, Britain and Spain ranging between 8,00,000 to 5,00,000 years have to be taken with a pinch of salt. Europe was going through the inter-glacial period of Gunz and Mindel. Homo Erectus was completely naked for all practical purposes without any hair or fur or drapery of animal skins to survive in such northern climates. We know he has mastered the use of fire by 4,00,000 years from the remains in Choukoutien caves. It is unlikely he mastered fire as long as 8,00,000 years ago.

---

* Rick Gore: *Expanding Worlds*: *National Geographic*, May, 1997.

* Rick Gore: *The First Europeans*: *National Geographic*, July, 1997.

What is more the dating techniques themselves have undergone a sea change after the war. University of California at Berkeley has pioneered radio active dating techniques which are much better than stratigraphic or geomagnetic techniques. For radioactive techniques to be of any use we need volcanic ash deposits which contain radioactive materials. The techniques used in Italy, Britain, Spain does not indicate any use of radioactive techniques. We can not take the older ages of 800 thousand years as a reliable date for European Homo Erectus fossils found there. They could have migrated there between those interglacial priods. They might have evolved into the subsequent Neandertal species as a result of the adaptation to colder climates.

Other important Homo Erectus fossils were found in Olduvai Gorge itself in bed II by LSB Leakey and Mary. Their son Richard and the American Alan Walker found a Homo Erectus skeleton of a 12 year old boy in 1984 near lake Turkana in northern Kenya. Other finds were made at Ternifine, North Africa and Vertesszolls Hungary.

All these finds range between 1.5 million to 4,00,000 years. Four marked differences stand out to differentiate them into a separate genus from those fossils of Australopithecus. They are (1) a much better erect posture almost on par with modern man, (2) a much enlarged brain capacity ranging from 775 c.c. to 1225 c.c with an average of 1000 c.c., (3) refinement of tools and materials used for tools and spread to Asia as well as Europe, (4) Mastery of fire.

We will discuss them one by one. By far the most important find was at the Choukoutien caves in China. The sheer size of the sediments which are 160 feet deep and the number of fossils found not only of Homo Erectus but also those animals he killed and ate are quite large compared to any other site. Above all undisputed evidence of the mastery of fire is found there. It is as important as the making of tools in furthering human evolution. Evidence of cannibalism is another important aspect which was found here which can not be ignored.

The 160 feet of sediments at Choukoutien were divided into several layers of occupation. The lowest layers showed signs of occupation by carnivores. The fossils belonged to them as well as their kills. The middle layers showed signs of occupation by carnivores alternating with those of Homo erectus occupation. This indicates sometimes the Homos drove out the carnivores and occupied the caves and at other times the carnivores drove out the Homos. Finally the top layers were completely occupied by Homos only. By this time signs of mastering fire also were evident. We could guess that the Homos drove out the carnivores finally with the help of fire.

The fossils include those of nearly 60 animals ranging from rodents and bats, beavers, water buffalloes, horses, deer, rhinoceroses and elephants to the sabertoothed tigers and huge hyenas. The Homo fossils include those of about 38 individuals and include fragments of jaws and skulls and some teeth. Only six skulls have the vaults intact. All have marks of violent blows. The bottom or back of the skulls were smashed as if for the brains inside. Apart from these, bones were practically few from other parts of the skeleton. The long bones were smashed and charred as if for the marrow. It appears that the Sinanthropus was a cannibal and ate his own kind. Whether it is a ritualistic eating or simply as a matter of another animal to be eaten is not known. If awareness and self-awareness had any dawning in Homo Erectus we should think that the cannibalism is ritualistic. We know from 19th and 20th century ethnographers that isolated societies do practice cannibalism and their enemies slain in war were eaten in the belief that their enemies' powers will vest with them. Other social groups eat their dead relatives in the belief that they live closer to them even after physical death. We know that chimpanzees exhibit some degree of awareness. It is quite possible that the Homo Erectus also had a degree of awareness and their cannibalism is ritualistic and not merely for food.

The other remarkable find of the Choukoutien caves is the evidence of mastery of fire by the Homo Erectus. The occupation layers in the top 18 feet contain small ash deposits here and there as well as charred bones and stones. This indicates mastery of fire by the Peking man. He used it for cooking, warmth and protection from predators. As with the making of tools so the control of fire is a major step which contributed in the cultural innovation as well as in the physical evolution of Homos. These two important events were separated by as much as 1.5 million years or more. Tool making and use was preceded by using tools which occurred naturally like stones, sticks or animal bones from carnivore kills. Then only tools were made to a design and the earliest design was called Oldowan and remained without much improvement for a long time. So also, perhaps the emerging Homo must have been observing naturally occuring fires like the forest fires due to lightning or friction of trees in summer months. Homos were living in bush country and savanna cuntry on the edges of forests. When he observes that a forest fire dies down when there is no vegetation to feed it, his best escape lies in running towards open country. Those unfortunate animals and birds which could not escape the fire are roasted alive. Homos are no doubt curious with that large brain and when they escaped the fire, the next step would be to pick up some embers from the dying fire and find out which end of the burning stick hurts and how to keep the fire going by feeding small amounts of sticks to it. Some of the roasted animals might also have been picked up for readymade food and when once the discovery is made that cooked meat is easier to skin and more palatable to eat, regular roasting of animals on a controlled fire is possible. Thus control of fire gave man warmth, cooked food and protection from predators by scaring them away. All this might have been repeated by many Homos many times before they were finally able to use fire by saving it from natural fires feeding it small amounts of fuel. When once a fire goes out they have to wait for the next natural fire. So the next step is that they have to learn how to make it themselves.* Clark Howell observes that the Homos could not have missed the spark that might fly off while making stone tools by striking one hard stone against another. Such sparks might have ignited the dry grass or leaves or fur of the animals which might be used as a cushion and thus early man might have learned to make fire himeself.

What is more that the Homos have used fire drives by igniting dry grass and thereby scaring elephants and other

---

* Clark Howell: *Early Man, Time-Life Books*, 1970.

large game to plunge into bogs and get themselves mired and get dispatched by the Homos at leisure. This requires cooperation between more than one group. Such cooperation also requires elaborate planning and some sort of advanced communication between the members.

Clark Howell excavated such a place in the Ambrona river valley in Spain at two sites, one near Ambrona town and another near the town of Torralba. There are clear evidences of tools, ash deposits and finally the bones of large elephants. The bones are not articulated with each other *i.e.,* in a position of natural joints but the large leg bones as well as a tusk were aligned in a line as if to form a causeway in the bog from dry land to the carcasses. All these evidences show us that Homo erectus by 400 thousand years was not only a master of more advanced stone tool techniques but mastered the making and using of fire.

The tools he used were an advance on the simple tools of Oldowan industry. The Homo erectus tools were hand axes and are known to belong to Chellean or Acheulian industry, after St. Acheul a place name in France where the tools were first found. These tools were found in many river valleys of Western Europe, Africa and parts of Asia since suitable pebbles for tool making are found in river beds. It is something like the availability of raw materials which determine the location of large industries like steel making in the 20th century. The spread of Homo Erectus also depended on the availability of suitable stones for tool making.

Basically the Acheulian tools are of two types. One is a core tool and the other is a flake tool. In the core tool the pebble is chipped away all round and a bifacial hand axe type tool is produced which could be used for heavy hacking. The flakes are those pieces which are the by products of the core tools. The flakes are more sharp and could be used for cutting. If lighter hammers are used made of wood or bone some finishing touches could be given to these tools to produce sharper edges by removing smaller flakes. This industry lasted for nearly 400 thousand years from 500 thousand to 100 thousand years ago. The earlier Oldowan industry persisted for nearly 1.5 million years from 2 million years to 500 thousand years ago. The evolution of Australopithecines also persisted for a long time. It looks as though that the tool making and other cultural abilities acted as a spur on the physical evolution of hominids. It could also be vice versa — the physical evolution spurred tool making and other cultural innovations and abilities.

It is a sort of feed back arrangement. One influencing the other. We see in the next stage of evolution — that of Neandertals and the Cro magnons a faster pace of cultural adaptation to suit varied environments and living requirements. That forms the subject matter of the next two chapters.

We will consider here one last feature of Homo Erectus. What is the level of his communication capabilities. Physical evidence like the development of brain areas associated with speech, the hyoid bone behind the tongue which anchors some tongue muscles, and the last vertebra of the thorax known as T7 which allows the passage of nerves that control fine breathing are not developed sufficiently to allow for articulate speech. Howerver communication through short sounds is not ruled out. What is required is capacity for conceptualisation. Tool making also requires conceptualisation. The tool had to be visualized in the 'mental eye' before it is made. When tools were visualized so, for nearly 1.5 million years and the tool making technique also had undergone a jump in technique from Oldown choppers to Acheulian we could assume that the conceptualising capacities also have vastly improved by 500-400 thousand years. At least they could have associated some definite sounds with some objects and other sounds with some actions is quite possible. If they acquired some mastery over some nouns and verbs it is quite sufficient to their way of life and activities to communicate with each other. The sounds in nature themselves provide some guidance as to what sounds to be associated with what objects or actions. Thus we could assume that Homo Erectus by 500-400 thousand years ago was able to communicate with each other not only through gestures and bodily postures only but with some sort of simple word association with objects and activities. This can not be compared with the speech of modern humans but is definitely an improvement than the mere body and facial expressions and sounds available to chimpanzees and other primates.

By and large Homo Erectus is a meat eater and hunter. The fossil remains of his kills ranged from mice to elephants. He did not even leave the brain and marrow. He ate everything. He had no compunction to eating his own kind. Whether it is for food or as a matter of ritual is not clear. Other carnivores do not eat brains or marrow of their prey, particularly antelopes because it is hard to chew through the bone. Homo Erectus with his stone tools did not have any such difficulty. He smashed the bones and scooped up the brains as well as marrow. This is a highly energy rich and nutritious food. This helped in the growth of his large brain. Pound for pound nerve tissue requires more nutrition than the tissue of any other body parts. We could even say that without a meat eating period in human evolution, the evolution of a large brain would have become impossible.

Meat not only supplied the energy required for the developing brain but the hunt of large animals required on the spot decision making ability as well as anticipating the animal's moves. That means the instinctive behaviour that is available to the animals is not sufficient. The plastic behaviour which collects information from the several sense organs like vision, hearing, touch etc, should be integrated in the brain and appropriate reactions should be passed on to the locomotor limbs of the body. This should be done moment by moment when one is involved in an active hunt of elusive small animals or dangerous large animals. That requires the evolution of a larger brain.

Not only that, such hunts are not possible for individuals. There has to be collective and cooperative effort. For that there has to be some social adhesiveness and organization between individuals. What better social adhesive is there than male female bond, their children and extended families. In such groups there has to be necessarily some kind of communication beyond mere bodily or facial postures and sounds. There has to be some association of some sounds with definite objects and actions and these should be understood by the group and passed on to posterity. All these activities necessarily act as selection pressures on the evolution of larger brains which can perform these related activities. Thus Homo Erectus with his brain of 900 c.c is an advance on Australopithecus.

We have Homo Erectus fossils dating from about 1.5 million years in lake Turkana area of Kenya to 600 thousand years fossils of Java and 400 thousand years fossils of China to 700-300 thousand fossils of European sites. After that there is a gap of nearly 200 thousand years between 300 thousand and 100 thousand years. Then suddenly by 100 thousand years we have a new type of Human known as the Neandertal man which is a much advanced homo than Homo Erectus.

There are however two skull specimens available from this long gap. They are dated somewhere around 230 to 250 thousand years ago. One is from the Swanscombe in the Thames valley of England and the second one is from Steinheim in Germany. The Swanscombe skull had only the upper and back portion of the skull whereas the Steinheim skull had also the face preserved. They are not primitive enough to be classified as Homo Erectus or advanced enough to be classified as Neandertals. Their features and range of variation is in between the two. We could say Homo Erectus is evolving to Neandertal type in the inter glacial conditions that prevailed in Europe at that time. Many of the Neandertal type of fossils also came from Europe.

10
NEANDERTALS

In 1848 a human type of skull was found in Gibraltar in a quarry on the north face of 'The Rock'. However its importance was not recognized for another half a century. It belongs to one of those type of skulls which we now call Neandertals. The honour went to the Neander valley in Germany near Dusseldorf, after which the Neandertal type of ancient man was named.

In 1856 workmen quarrying for lime stone in the Neander valley found the skull and a few rib and limb bones of a human. The bones reached ultimately to Prof. Hans Virchow Von Holder the leading anatomist of Germany of those times. He however dismissed the fossils as belonging to an idiot suffering from malnutrition and rickets and arthritis. No other animal bones or artefacts were found along with the fossils.

The English geologist Sri Charles Lyell and biologist Thomas Huxley who studied the cast of the Neandertal skull recognised it for what it is, a variant form of man with some apish features. Prof. William King of Ireland named it Homo Neanderthalensis in recognition of its relationship to man.

In 1886 geologists from Leize University Belgium found the remains of two human skeletons in a cave called Grotto de Spy. The skeletons consisted of two brain cases, two lower jaws and some other bones. They exhibited the same features as the Neandertal skull. Low skull vault, receding brow and massive brow ridges over the eyes are the characteristics of Neandertals. These are the same features we observed in the case of Homo erectus also. However the Neandertals have more brain capacity, almost equal to that of modern man.

The brain case is broad and long at the base. It gives him a bull necked appearance. In the Spy cave are also found Paleolithic stone tools as well as the remains of cold climate animals like woolly rhinoceros, woolly mammoth, cave bear and cave hyena.

A cave near Le Moustier in the Dordogne region of France was investigated by Edouard Lartet a prehistorian of France. The implements found there and similar implements found at other prehistoric sites were hence forward regarded as belonging to Mousterian culture. They were made of unpolished stone and many of them were trimmed on one side. Now it was found that the implements found at the Spy in Belgium along with the two Neandertal type skulls also resembled those of Mousterian culture.

At Le Moustier itself no fossils of ancient man were discovered. So at that time the Mousterian culture was not identified with any particular type of ancient man. With Spy discoveries Neandertalers were linked together with the Mousterian culture. It was also verified by latter discoveries.

In 1909 Otto Hauser of Switzerland dug up at Le Moustier in the Dordogne valley on the banks of River Vezere the grave of a 16 year old Neandertal boy. It showed signs of ritual burial and Mousterian type implements were placed near the boy. This unmistakably identified the association of Mousterian culture with Neandertalers.

In 1907 a lower jaw bone was found in the sand pits of Mauer near the town of Heidelberg Germany. It had all the teeth intact. It was thick and robust and chinless. However the teeth were ordinary. Heidelberg jaw might belong to a transitory form of homo between Steinheim man and Neandertal man according to experts.

In 1908 three French Priests found some cold climate fauna and Mousterian type of artefacts in a cave associated with a Neandertal type burial. The cave was near the village of La Chapelle-aux-Saints in Southern France. The cold climate fauna are woolley rhinoceros, extinct bison called aurochs and reindeer. Here in a shallow depression they found a Neandertal type skeleton laid fully extended on its back and head towards east. Besides it are Mousterian type artefacts and fragments of red ochre suggesting a ritual burial. Along with the skeleton were also found several limb bones of a bison suggesting that the corpse was provided with food also for its forward journey after death. This suggests that the Neandertalers developed some kind of belief in life after death.

In 1909 excavations at La Ferrassi revealed further human bones under a rock shelter. They were also associated with Mousterian type implements. The remains of a female and three children were also found here.

In 1899 at a place called Krapina in Croatia a number of Neandertal type skeletons were found. There were at least 10 of them of both sexes and of several ages. There were atleast 9 layers of human occupation at this site. Many of the human bones were broken open and charred by fire. It was however not known whether they were eaten as a matter of food or as a ritual. Since we find evidence for ceremonial burials, we could assume the cannibalism was also ceremonial.

So far we are describing the classic type of Neandertals that lived in the lime stone rock shelters and caves of France, Belgium, Germany, Ugoslavia, Spain and Italy. Their dwellings are full of human occupation signs like hearths, animal bones of heavy and fur covered woolly elephant, woolly rhinoceros, cave bears and aurochs an extinct type of big bulls. All these are dangerous animals. Their implements and weapons were mostly Mousterian. No throwing implements or weapons like arrows were found. They could have sharpened wooden spears with scrapers and hardened them with fire. This type of close encounter hunting of huge and dangerous animals require powerful bodies for the hunters too. From the evidence of the limb bones of the Neandertalers recovered it is evident that they were built quite powerfully and their average height is 5'.0 feet to 5'.3 inches.

The brain case of these classic type of Neandertalers is long and the cranium low and flat. The brow ridges are massive and the brow receding. There is an occipital bun like growth of bone at the back of the skull perhaps to counterbalance the long and protruding face and jaws. The eye sockets are large, the nose is long and broad, the upper jaw is protruding, the lower jaw strong and chinless. The bone of the brain case is quite thick and the brain capacity is generally considered to be on par with modern man, an average of 1300 c.c. The ribs are heavy and the rib cage large. The thigh bones are a little bent giving them a little bow legged appearance.

These classic Neandertalers lived in Europe when Europe was passing through its last glaciation the Wurm and the inter glacial period of Riss-Wurm. The time is generally estimated at 120 to 35 thousand years ago. Their special racial features were evolved perhaps as a result of the severe cold conditions under which they had to live and the heavy and dangerous animals, they had to hunt for their survival. No doubt animal's skins might have served as loose draperies though not shaped into clothes. Fire certainly helped them to protect from the cold and cook their food.

Thus by 1910 it had become evident that the Neandertalers were an ancient type of humans who were somewhere between Homo erectus and the modern man represented by the latter type of humans known as Cro-Magnons. Most of these early or classic type of Neandertalers were from Western Europe and from a time period ranging from 230 thousand years to 35 thousand years. The northern parts of Europe, Russia and North American Continent were going through the last interglacial period of Riss-Wurm and the glacial period of Wurm (70-10 thousand) at that time.

Most of the excavations conducted were by Europeans in Europe and the type of Neandertalers found was one variation or racial type of Neandertalers, the type which is adapted to a cold climate existence. However this is not the only type of Neandertalers found. This was known as the classical type and it has disappeared abruptly by 35 thousand years ago and the caves and rock shelters of Western Europe came to be occupied by another type of man, the Cro-Magnon man described in the next chapter.

Other types of Neandertalers came to light at places like Broken hill in a Tin and Zinc mine in Rhodesia (now Zimbabwe) in 1921, and the type of skull is known as Rhodesian man. The skull capacity is high, the brow ridges massive, the face very long.

During the period between 1931 and 41 nearly 10 skulls

were dug up at a place called Ngandong on Solo River in Java. The skulls have massive brow ridges, lacked faces and the top of the crania fractured as if to get at the brains inside. The skull capacity varied between 1150 and 1300. Weidenreich who studied them ascribed them to a variant of Neandertalers.

Thus we see Neandertals was a widely distributed species of the evolving man. Since the first remains almost all came from Western Europe and they were of a massive type, an opinion prevailed that the Neandertals were short and brutish fellows with crude skills and little refinement in implements or art and lacked any other finer feelings. At the same time the Neandertal remains were coming to light in France since the middle of 19th century, another type of human remains which were called the Cro-Magnons also were dug up almost in the same areas in France as the Neandertal remains were dug up. These Cro-Magnons were exquisite artists, superb craftsmen and exhibited a degree of concern to their dead that equals moderns. Their skeletal structures were also similar to those of modern man. All these developments have prompted the scientists to conclude that the classic Neandertalers of Europe were an extinct side branch of mankind and had no further role in the evolution of man. This was the situation until 1950.

However since 1950 and subsequently it is realized that the Neandertalers were as varied as the modern humans with many racial variations. This view was forced upon the scientists by some important finds in Palestine and Iraq, Gibralter, Croatia etc. The Gibralter and Croatian skulls were less massive in their brow ridges giving the first indication that the Neandertalers occupying the southern and more temperate regions were evolving in a different way than their northern cousins. They are less massive, taller and slimmer.

The skeleton of a woman with these finer features was found in a cave at Mount Carmel in Palestine in 1932. She had less massive brow ridges and the back of the skull was more rounded whereas the classic European Neandertalers had an occipital bun. This skeleton was known as the Tabun Woman. In a nearby rock shelter at Skhul nearly 10 skeletons were found and they ranged in variety from classic Neandertal to scarcely Neandertal at all. Such racial variation at the same place is rather an astonishing feature.

Starting in 1937 a cave near Shanidar in northern Iraq was dug up and ultimately the skeletons of 6 adults and a child were found. The skeletons were aged at 44,000 years. The features of their skull resembled those of Tabun woman. The brow ridges were less massive. The occipital bun less pronounced the cranium of the skull less flattened. They exhibit features of a transitory race between Neandertalers and modern man. What is more, the skeletons of a male, a female and the child show pollen of spring flowers round their skeletons indicating reverential burial.

These southern Neandertalers are known as modern Neandertalers. They in all probability moved down to the south from the northern climates as the ice sheets descended down over northern Europe around 70 thousand years ago. Those who remained in caves and rock shelters of France adapted to the cold climate prevailing in Western Europe and developed short statures with barrel chests, short and massive limb bones with slightly bowed leg and arm bones with heavy joints as well as the prognathous jaws, heavy brow ridges, receding brow, flat cranium, occipital bun etc., which were the characteristics of classical Neandertalers. They suddenly disappeared from the caves of Europe around 35,000 years ago. Skeletons of other people whom we now call Cro-Magnons began to appear. The classic Neandertalers might have become extinct unable to compete with the better evolved Cro-Magnons with their more sophisticated tool kits, their cold adapted sewed dresses and ability to build shelters and ability to build rock lined hearths, all of which indicate that the Cro-Magnons gradually edged out the Neandertalers in the Western Europe.

The story indicted by the southern and middle Eastern Neandertalers as shown by Tabun, Shukl, Shanidar specimens is a different one. There is a gradual evolution from Neandertal to Cro-Magnon type of people or even a blending of the races if Cro-Magnon type of ancestors moved up from Africa to meet the Neandertalers moving down from Europe to meet in the middle East and then blended together exchanging genes. In that case we need to find out Cro-Magnon type of people evolving somewhere in Africa some time between 200 thousand to 100 thousand years. The African prehistory is a blank during this period. So we have to rest content for the present that the Cro-Magnons evolved somewhere in the middle East and moved North into Europe to displace the Neandertalers.

We have to make an assessment of Neandertalers now and their place in human prehistory. Neandertalers were widely distributed hominids who also exhibited wide variation of a racial nature. They ranged from the glacial colds of Europe to the temperate zones of southern Europe, mediterranean coast, North Africa, middle East, Sub Saharan Africa to the East Indian islands. These are the places where we found their remains. They certainly could have ranged in more areas but we do not have any fossils or tools used by them from other areas.

The tools associated with Neandertalers were first found at a place called Le Moustier in the Dordogne region of Southern France and since then they were called the Mousterian culture. This culture had more variety of tools than the previous Chellean Acheulean cultures which were used by homo Erectus. In the Mousterian culture the flakes were more extensively shaped into different kinds of implements like scrapers, points, by detaching smaller flakes by applying pressure rather than blows. This produces a sharper edge. Thus, sharp points or knife like edges are produced. A third type is a sharp convex edge that could be used as planes to sharpen wooden spears. Many types of points were made that could be attached to shafts of wood that could be used as spears in hunting. Francois Bordes* professor of prehistory at the University of Bordeaux, France excavated a cave at Combe Grenal which contained at least 64 levels of occupation during a span of 85,000 years. It was a site of Neandertal occupation which contained many of the tools he made and used like spear points, scrapers, points, cores etc. At La Quina three different strata of Mousterian Culture were

* F. Clark Howell: *Early Man*, pp. 137, 139, Time-Life International.

unearthed yielding a variety of stone objects like awls, scrapers, saws, lance heads etc.

With these relatively simple tool kits Neandertalers managed quite well both during the temperate as well as during cold climates. He hunted down surprisingly large animals like woolly rhinoceros, woolly mammoth, cave bear and cave hyena all cold climate animals as evidenced at Spy. At La chapelle-aux-saints the remains of extinct bison and reindeer were also found apart from the other. Other sites indicate that he killed and ate a great many horses, elk and ibexes also. All these animals are large, wily, swift or dangerous.

He drove some of them like the ibexes or horses down steep cliffs into the valleys below where waiting hunters killed the wounded animals. Larger animals like the woolly rhinoceroses and woolly mammoths or the aurochs were hunted by driving them into bogs or pits and then thrusting spears into them.

The Neandertalers never developed throwing weapons as the latter Cro-Magnons did. This involved them in close encounters with often dangerous and large animals. In this they often received serious injuries. Rick Gore* tells us about the shattering injuries suffered by many of the Neandertalers, probably in close encounter hunts of large and dangerous animals. Most of them suffered from severe arthritis also. Surprisingly enough, these injured people also showed signs of healing. Obviously some of the group had fed them and taken care of them.

The hunts of large and swift animals require group hunts, the members co-operating with each other planning the hunt in advance. The care taking of the injured shows that the members of the group are bound by more stronger bonds than mere cooperative hunting. They might be family bonds. The living in caves and rock overhangs during the Wurm glaciation indicates that the hunters had a home base to which they could return after the hunt is over. There the females could raise the children and scrape and clean the hides of the animals killed, for use as beddings or as simple draperies over the bodies. Many caves also show several feet

* Rick Gore: *Neandertals: National Geographic*, January 1996.

of ash deposits showing that the Neandertalers had a sure control of fire, making it at will either by friction of hard woods against one another or by striking one flint against another and feeding the sparks to flammable materials. Also the several animal bones found in the caves were charred and broken as if they were roasted and the marrow was scooped out. The skulls of Neandertal man found at Ehringsdorf in Germany, Krapina in Croatia and Solo in Java were broken open and the long bones broken and charred as if for the brains and marrow. Could we simply say it is an act of cannibalism. Obviously not. We have already discussed the theory of cannibalism in the last chapter dealing with Homo Erectus and found ourselves differing from the view of cannibalism practiced to appease hunger. There is all the more reason in the case of Neandertalers to dismiss the theory of cannibalism and adopt the theory of ritualistic eating of their own kind either probably enemies slain or loved ones honored in this way.

The Neandertalers were burying their dead in a ritualistic ceremony. They are carefully laying them down in sleeping positions. Fine Mousterian type implements were laid along side the bodies. The red iron oxide mixed clay of ochre was daubed on many of these dead bodies. Even spring time flowers were showered round some bodies. What is more, choice portions of bison were laid across the dead as if to provide food for them in their life after death. All these acts unmistakably indicate that the Neandertalers had some kind of belief for life after death. That indicates awareness and self awareness. We are not much different from them in these respects. We may be more sophisticated in our beliefs. But we too entertain beliefs about life after death. Under such circumstances it is not tenable to think that the Neandertalers simply ate their kind as a matter of food. Some kind of ritual is decidedly involved in their practice of cannibalism. What it is we can never know.

The burial practices also indicated some animal cult practices, particularly cave bear cults. Cave bears are huge and dangerous animals. Yet Neandertalers hunted them down and preserved their skulls. In Eastern Austria seven bear skulls were preserved in a pit. Again such a preservation was found in a stone lined pit in Drachenloch, Switzerland. In Southern France the remains of 21 cave bears were preserved in a pit and covered by a stone slab. In Uzbekistan goat skulls were arranged round a burial pit of a child. These practices undoubtedly point out towards some ritual significance.

As the ancient man developed more cognitive powers abstraction and symbolization which go hand in hand with the development of his memory power, its recall and image formation he began to have an awareness of himself and of the outside world and how much this relationship effected his welfare and how much he depended on it for his continued well being and survival. For this relationship to be smooth and balanced he developed ritual and cults which helped to regulate the behaviour of the members of his group towards each other as well as towards the outside world.

In a strict sense the ritual and the objects used for it are non-utilitarian. In a place near Tata in Hungary was found a section of a baby mammoth tooth carefully bevelled all around to an oval shape and then polished on both sides. This object is obviously of no utilitarian significance.

Yet the maker had devoted many hours of skillful craftsmanship to fashion it. Such objects together with the careful burials with stone implements, flowers, wreaths, animal meat together with evidence for the care bestowed on injured and sick persons of the group indicate a dawning of what we may call some of the essential characters of how humans differ from animals. The unique features of human evolution are not eating, procreating and dying but how they go about doing them. This sets them on the road of culture and civilization.

These capabilities definitely require a degree of communicative skills between the members of the group beyond mere bodily gestures and a few sounds. Now sounds should be produced at will and to a set pattern and these sounds are associated with concepts and understood as such by others. These concepts include symbolization of nouns, verbs and perhaps adjectives and adverbs as well as prepositions, gender and tense and number also. In their daily encounters with others as well as the animals, birds, trees and plants and grasses a distinction and differentiation between them and others was made and such awareness impose on the evolving hominids the selection pressure required to distinguish between the various differences of what we now call nouns, verbs, adjectives, adverbs, prepositions, gender and tense and number.

We can not know their development unless we have some tangible physical proof. The brain capacity of the Neandertalers was as large as Cro-Magnons. Only the shape is different. So the internal brain connections and relative development of the areas of brain controlling the various functions may also be different. The difference is most evident in the frontal lobe region. The Cro-Magnon brow is large, rounded, high and quite pronounced, whereas the Neandertal brow is low and flat and receding. Herein lies the chief difference. Also the chief difference between the latter Cro-Magnons and the Neandertalers was in the levels of abstract thinking.

What is of interest to us here is that the Neandertalers possessed in their skulls a small bone called the hyoid bone. In modern humans the voice box hangs from this bone at the back of the tongue. The controlled breath emanating from this voice box is further shaped by tongue and lips to produce the various sounds we now use for our speech. Since the Neandertalers possessed this bony structure almost similar to modern man, we could assume that they had a developed voice box which could produce the necessary sounds. What sounds were associated with what meanings was a matter which shows the level of abstract thinking reached by those homos. In this we could safely assume that the Neandertalers possessed simple language which could distinguish like different animals, places or other objects by different names, their actions by simple verbs and qualities by adverbs and adjectives, their relative positions by prepositions, their sex and age by gender words of distinction and their numbers by words for numbers. This level of abstraction, symbolization and word association are definitely within the reach of capabilities of Neandertalers.

From a cursory dismissal of Neandertalers as uncouth idiots suffering from rickets, a hundred and fifty years ago;

by the turn of 20th century they have received a respectable pedigree as to deserve a place in human ancestry and not an aberrant side branch which became extinct in the cold caves of Western Europe in the middle of Wurm glaciation.

11

CRO-MAGNONS

Edouard Lartet the noted prehistorian of 19th century France investigated some caves in the Dordogne valley of South Western France. Out of them two caves are of importance so far as they presented the animal remains and cultural artefacts associated with a type of people latter came to be known as Cro-Magnons. The skeletal remains of these people themselves were not found at these caves.

At the cave in Aurignac the remains of extinct cave bears, cave lions, mammoths, woolly rhinoceroses, bison and elk and reindeer were found. Many of the bones were charred and broken open for marrow. The fire indicated roasting of the meat.

At the cave near Le Medeleine, tools and weapons made of bones, antler and ivory were found. Also found was a thin plate of ivory on which was etched the outlines of an elephant with the long hair of a mammoth. The cultures associated with these cave finds were called the Aurignacian and Magdalenian. These finds were made between the years 1860 and 1864.

In 1865 at a place called Solutre, the remains of another culture was found which was since named Solutrean and most of the stone tools were shaped like laurel leaves. The animals slain for food were mostly horses. Their bones were charred by fire and broken open for marrow. The other tools of ivory and bone and stone as well as the other animal remains were similar to those found at the other sites. Yet no homo skeletal remains were found.

Then in 1868 during the widening of a road, a cave was opened up at a place known as Cro-Magnon in the Vazere vally of Dordogne. Louis Lartet, son of Edouard, investigated the cave and found the skeletal remains of five individuals. They belonged to an old man, two adult men, a woman and an unborn infant. They were found in association with the same type of extinct animals as well as the same type of cultural artefacts found at Aurignac. These were the people responsible for the Aurignacian, the Solutrean and the Magdalenian cultures. They came to be known as the Cro-Magnons since the first skeletal remains of this type of people were found at Cro-Magnon cave.

The skeletons of Cro-Magnons were quite different when compared with those of the Neandertal people, who also occupied the same Dordogne region of France during an earlier phase of the Wurm glaciation between 1,10,000 to 37,000 years ago. The skulls of the Cro-Magnons were quite massive and rounded. The brow was prominent and high. There is no occipital bun as in Neandertals. The brow ridges over the eyes were thin and not joined by a bar. The chin was marked and protruding. The jaws and teeth were lighter and not jutting forward as in Neandertals. The ribs were less heavy and the rib cage is smaller. The thigh bones were straight and long compared to the upper arm bones. The Neandertal thigh bones, leg and hand bones are more heavy and the thigh bone is bent giving the owner a bow legged appearance. The arch of the foot is more curved in Cro-Magnons than in Neandertals. In all these respects the contrast between Neandertals and the Cro-Magnons was quite remarkable. The Cro-Magnon skeletons were almost similar to the skeletons of modern men.

Further excavations at Grimaldi in the Southern French coast brought to light six skeletal remains associated with Aurignacian and Magdalenian cultures. Ochre was found near the skeletons indicating they were given some kind of ceremonial burial. Ornaments of shells and bones as well as tools and weapons were also found along with some skeletons. Some skeletons were found in such a position with their knees touching the chin and hands folded under the skull as if they were put in sleeping positions or foetal positions. All these indicate unmistakable signs of ceremonial burials.

The old stone Age or the Palaeolithic spanned nearly 2 million years. This period is known as the Pleistocene in Geological terms. This is further divided into three periods, the upper, the middle and the lower. The upper period is generally treated as starting at 1,10,000 years B.P. (Before Present) and the middle period at about 7,00,000 years ago. The lower period was roughly put at 2 million years. During this Pleistocene period as a whole important evolutionary developments took place in homos as we have seen in the last three chapters. They were as a result of the adaptations made by the homos to the changing environments in which they lived.

There were important climatic and ecological changes during this period. Vast ice sheets advanced and retreated over much of North America and Europe and glaciers in Africa as well as Europe and over the Alps. These have altered the shore lines of continents, though the continents themselves did not shift markedly. Due to the locking up of vast quantities of water in these ice sheets which are sometimes several hundred feet thick, the sea level was estimated to have fallen by 300 feet. This exposed the shallow seas and the continents were joined by land masses at several places. Thus the Indonesian islands were joined to mainland Asia, Tasmania and New Guinea were joined to Australia. Russia and America were joined at the Bering Strait, Britain was joined to Scandinavia. The shore lines of every continent extended much further into the sea than at present.

There were at least 15 such advances and retreats of the ice sheet during the last 700 thousand years. Not much is known about the ecology during the lower Pleistocene *i.e.,* between 2 million and 700 thousand years. Glaciations in Europe during the middle and upper pleistocene *i.e.,* during the last 700 thousand years were well understood and great ecological changes took place during this period not only in Europe but all over the world. As the glaciation advanced southwards the tundra extended southwards and so did the arid zones and the rain forests moved closer to the equator. When the ice sheet retreated, the reverse happened. The forest spread northwards, and so did the arid zone and the tundra. These shifts also shifted the animals and plants that are adapted to these zones. Thus these ecological changes effected the plant and animal life. Some species became extinct and others replaced them. Some species evolved to the changing ecological conditions. Homos who depended on plant and animal life directly or indirectly also have to move along with their food sources, evolve or perish. We have evidence for all three developments.

The early homos of Africa moved to south-east Asia, North Africa, China and even Europe by 700 thousand years. They have evolved too to Homo Erectus. Some species like the Zinjanthropus and Paranthropus became extinct. We also have evidence that the Neandertal variety of homos became extinct by 37,000 years B.P. We do not know for sure whether it is a species variation from Cro-Magnons or a racial variation of Cro-Magnons. Whatever is the case Cro-Magnons ruled the Earth since 37,000 years B.P.

The Cro-Magnons hunted both the Tundra animals like the reindeer, woolly mammoth, woolly rhinoceros, giant elk, and saiga antelope besides a number of small animals like arctic hare, lemmings, voles, and birds including geese and snowy patridge. Wild horse was a particularly favoured food animal. Further south in the Iberian peninsula and the Balkans and Italy where the temperate conditions favoured forest ecology, the animals hunted were red deer, aurochs and wild boar. The heaps of animal bones both in caves and in open country together with the tools and implements found there testify to these animal hunts.

As noted already there were several oscillations of the ice sheet and along with it the ecological conditions also changed. These changes required different cultural adaptations by the Cro-Magnons since they depended now more on cultural adaptations than on physical evolution. There are three principal cultural adaptations. They are the Aurignacian, the Solutrean and the Magdalenean. The Aurignacian had sub phases, the important ones being the Gravettian and the Perigordian. The Auriganacian culture lasted between 32-27 thousand years B.P.

The Gravettian lasted between 26-20 thousand years. The Perigordian lasted between 20-18 thousand years. The Solutrean lasted between 18-16 thousand years and the Magdalenean between 15-8 thousand years. After 8 thousand years B.C. the ice sheets retreated and the Holecene period or the present period followed. The present type of atmosphere and seasonal conditions and variations, together with the flora and fauna of the present day evolved. This resulted in the adaptations of entirely different cultures by the Cro-Magnons.

The Cro-Magnon tool kits were specific to suit a particular need. They also varied for the various seasonal conditions and purposes to which they were put. The general tool was a long blade. It is rather an intermediary product to make specific tools. It was long and thin. A fine blade of this type was 13 inches long and ¼ inch wide. In this technique of tool making a cylindrical core was first prepared out of flint by striking off rough edges. Then using a punch and hammer, blades were struck off its cylindrical face. Then specific tools for specific purposes were struck out of this blade by a method known as pressure flaking, A bone or antler hammer is used to flake smaller bits of flint. Applying pressure flaking on one longitudinal side of the blade, a sharper blade could be obtained. Fine flakes could be removed from one end to obtain convex or concave scrapers. By striking off a sharp edge from one end a sharp cutting edge could be obtained called a burin, which could be used to work softer materials like bone, antler or ivory. By cutting two sharp edges on either side a needle like tool could be obtained to work on leather.

In this method of tool making material was less wasteful. Nearly 75 feet of cutting edges could be obtained from a two pound flint stone rather than the 6 ft of cutting edge obtained by the earlier Mousterian technique used by Neandertals. As noted earlier the blade itself is not the end product but is an intermediary piece out of which the end pieces were made. The tool maker had to visualise the end product in his imagination and plan the sequence of operations required. This requires in the Cro-Magnons a degree of conceptualisation which is much more advanced than is exhibited by the Neandertals. This necessarily implies an enlargement of the frontal cortex and the corresponding changes in the shape and size of the skull.

The Aurignacian culture was followed by the Gravettian and the Perigordian cultures. A milder temperate climate prompted them and the Cro-Magnons who created these cultures lived more on the lake shores and river banks in the open country extended between the Pyrenees and the Ural mountains. Their settlements are found in France, Central Europe and Southern Russia. They lived sometimes in round huts with the earth scooped out to make a pit, the roof and sides were supported by animal skins weighed down by mammoth tusks or long bones. Light tree branches may also have been used for the roof supported by light posts. However there is no evidence to support the use of heavy timbers. No implements that are needed to work on heavy timbers were found. Their culture and economy depended on herd animals of the tundra and animal products like skins, bones, ivory, antlers sinews etc. These were cut to proper sizes with flint burins and ground and polished to make sharp points to be fitted to spears and other throwing weapons. We also find burins and needles that could be used to work skins and in all probability they made and wore skin clothing to protect from cold. They also used ornaments of shells, animal teeth and pendants made out of ivory, sometimes carved with intricate designs.

Apart from the clothing and huts made by the people of Gravettian and Perigordian cultures, they also made the remarkable venus figurines. These figurines are found from France to Central Europe to Southern Russia. The figurines were only a few inches tall and were made out of ivory, various stones, or a mixture of bone dust and clay and fired for hardness. The figurines were so named because of the exaggerated importance given to the breasts, hips and belly in other words to the feminine features. They suggested pregnancy and fertility symbols. The hands and legs and even the head were merely suggested without any clear shape and quite often diminutive in size compared to the fertility symbols. The extensive area in which they were found extending from France to Southern Russia indicate that these people shared a common psychic life and belief systems. What is their belief system, we can only guess.

The Gravettian culture was followed by the Solutrean in parts of France and Spain. These people are regarded as immigrants from the East perhaps from Hungary. They lived in the open perhaps in a temperate climate. The stone tool technique was further refined by fire treating the flint cores. Such cores are easier to cleave long blades from the sides and they are softer to work with and feel greasy to the touch. Fine pressure flaking of these blades produced laurel leaf shaped double edged blades and arrow points with single or double shoulders. At a place known as Solutre in France, bones of nearly 10,000 wild horses were found which were eaten by these people. These people also made a number of decorative articles of a non-utilitarian variety out of pieces of mammoth ivory, the teeth of arctic fox, wolf and sometimes even of cave bears, shells and cowries and ground stones. Decorative designs were also carved on ivory, antler and bone.

There was again some climatic change starting around 17,000 years and with it there were changes and improvements in the cultural artefacts also. This period between 17,000 to 8,000 B.C. is known as Magdalenian. They produced a large variety of fine tools for hunting like harpoons, fish hooks, arrow heads, spears and needles. The sharp instruments made out of bone, antler etc. that could be fitted to throwing weapons was the main feature of this culture. Bone needles with an eye as well as fish hooks have also made their appearance. That indicates they made and wore animal skin clothes. A figurine of a male carved in ivory was found in Russia. It wore a hood made of fur as well as clothes, trousers and boots, made out of animal furs.

This clearly indicates that these people wore clothes like the present day Eskimos. They also caught fish most probably in the rivers when the first thaw occurred in the spring and salmon started swimming up river. It is however for the remarkable cave art that the Magdalenian culture is remembered. Before that one word about the ceremonial burial practices of the Cro-Magnons.

We have seen in the last chapter that there was evidence of some sort of burial given to the dead of the Neandertals. Also red ochre, pollen of flowers and some tools and weapons and sometimes animal bones were also found near the burials. These burials give us an idea of the psychic life of these people. In the case of Cro-Magnons almost all the discoveries of their skeletons showed unmistakable signs of ceremonial burial. The early discoveries made in the second half of 19th century and the early part of 20th century were not done with a professional thoroughness to dispell doubts about upper Paleolithic men giving ceremonial burials to their dead.

Thus in 1872 a human skeleton was found in the rocks of Grimaldi in Southern French Coast. It was stained with red ochre and surrounded with fauna and cultural objects of the Aurignacian Period. The position of the skeleton also indicated that the person was given some sort of burial. In 1894 the skeleton of a youth was found again in southern France. The skeleton was covered with red ochre and the cultural implements found along with it belonged to Magdalenian Culture.

Prince Albert of Monaco organized some excavations near the Grimaldi caves by the end of 19th Century. He secured the help of experts of France of the day in the required disciplines. The scientists unearthed four human skeletons from three different occupational levels. All four bodies were carefully buried. Each was stained with red ochre and surrounded by ornaments. The two skeletons from the upper layer were fully extended and the two in the lower layer were placed in a contracted position. These carefully observed discoveries by experts in the field clearly established that the Cro-Magnons gave a ceremonial burial to their dead.

These ceremonial burials were by no means restricted to the caves of Southern France. Fortuitous conditions preserved them. These burials were extended throughout Europe where Cro-Magnons flourished. Grahame Clark* writes about a carefully excavated burial at a place called Sungir near Moscow. A 60 year old man and two boys were buried together in an

* Grahame Clark: *World Prehistory*, p.104, Cambridge University Press, Cambridge.

extended position. They were daubed with red ochre. They seem to be fully dressed when buried from the disposition of nearly 3000 beads which might have been attached to the clothing. The ornaments are of perforated animal teeth, ground stone pebbles and mammoth ivory beads and roundels.

Personal ornaments from other parts of Europe include fish vertebrae perforated snail shells, Arctic fox teeth and other animal teeth. A variety of beads, pendants, pins and bracelets were also made which were also further decorated by carved patterns. Personal decoration as well as using paints and dress are common among these people who lived over an extensive area from the Pyrenees to the Urals in Russia. They seem to have shared a common psychic life as well as a social pattern based on their common belief system. Their art and technology also exhibit similarities. For the first time we find that these people are spending more time and effort on non utilitarian activities than on hunting and gathering activities. This means that they are creating an order out of their experience and awareness. This could be better understood in their art which we will examine now.

Art made its appearance for the first time with the Cro-Magnons even in its earliest cultural phase, the Aurignacian. It really flourished in the next Gravettian period which was marked by the venus figurines discussed earlier. Apart from the Venus figurines they also made clay models of a variety of animals and sculpted lime stone reliefs of bison, horse, and ibex as well as mammoth, rhinoceros, reindeer, lion and tiger, wolf and lynx. These figurines were found extensively in the area between the Pyrenees and the Urals.

The best Art however came from the Magdalenian Period. This includes engraved, carved and painted animal figures on rock surfaces, usually in caves. This extended from southern Spain to the Urals wherever nature provided suitable caves. It is however concentrated in Spain and France where it was first found out at Altamira in Spain and at Lascaux in the Dordogne valley of France. There are many other sites of cave art from southern Spain to the lower Rhone basin and the southern parts of Italy. In the southern Ural mountains also in the southern bend of Byelaye river there are cave paintings resembling those in Dordogne. This shows that the community of upper Palaeolithic people was quite extensive.

The art depicted on the walls of caves was mostly of herd animals like those the Cro-Magnons hunted for their food, clothing, shelter, tools and ornaments and other needs. The caves were often deep and dark and inaccessible. There were no indications that the caves ever served as living spaces. The art itself was in dark niches as if to indicate that it was not meant for display or appreciation. Many animals were depicted as in advanced stage of pregnancy. Some also bore symbolic arrows and blows with clubs or stones. Many a time animals were painted one over the other showing that a particular wall face was a favoured spot for painting than others. The natural shapes and contours of the cave walls were also taken advantage of to depict the animals in more realistic perspective. The painters were undoubtedly masters at their work. The colours used were various shades of iron oxide ranging from chacolate to brown to orange and red known generally as ochre. The black colours were provided by Aluminium Oxides.

What was the purpose of this art then? Experts are of the opinion that this art was symbolic to indicate sympathetic magic. By painting the animal on the wall and performing certain rituals, the hunter expects to gain power over the animal which will submit easily to its being hunted. That is perhaps the reason why all these caves never contained a human figure painted on the walls. If they do, some hunter would also lose life as the animal painted on the walls. These paintings are symbolic and the represented animals are the hunted animals. Hunting ritual and magic is involved with it. The caves also depict paintings of persons donning animal head dress and skins depicted in some kind of ritual dance. These figures are interpreted as those of the magicians or even as the gods of the hunt. In all probability these magician figures are the ones who painted the animal and other figures which required a high level of proficiency and observational skills as these animal figures are quite realistic in many cases. These shamans then used the caves as the places of ritual for the community which they served. There are some other interpretations also, notably these animals are treated as totems, the ancestral animals from which those people descended. Others interpreted them as sex symbols, some species representing male principle and some others female principle.

Alexander Marshak* not satisfied with these traditional interpretations of Cro-Magnon art, tried to reinterpret them using modern techniques of study and analysis. He subjected the artefacts to microscopic examination and photographed the wall paintings under infrared and ultraviolet light. He writes about two artefacts which he examined under microscope and which yielded significant clues to the mind of Ice Age Man. One is a reindeer bone aged about 30,000 years and four inches long. From the wear marks on it he deduced it was used for pressure flaking flint blades. On one face of its surface there were 69 markings gouged with sharp instruments showing some kind of notation. Marshak thinks it was some kind of a sequence of events over a period of 2¼ lunar months.

Another artefact of which Marshak writes is a reindeer antler about 14 inches long on which were engraved a number of figures. First there were two seals one male and the other female. There is a salmon near them upside down with the characteristic curved lower jaw which males develop during the salmon mating season in the spring rush upstream of rivers to the spawning ground. Seals follow them to gorge on the salmon. Next there are two snakes with their genitalia showing as in the spring mating season. Next there are some wild grasses which bore coarse grain eaten in the ancient days. A sprout and a flower and three many legged damp soil creatures are also engraved on the antler. Lastly Marshak talks about a schematic ibex head which was crossed over. The other figures were quite realistic whereas the ibex figure was not. Marshak thinks the whole as a seasonal composition heralding spring and the schematic ibex figure was a symbolic sacrifice to it.

He also writes about a bone ceremonial knife found at La Vache. On the two faces of the knife two seasonal compositions were engraved. On one side were engraved the head of a doe

* Alexander Marshak, *National Geographic,* p. 62, January 1975.

and three lines which he interprets as symbolic for flowing water, three plants and an ibex head one of its horns crossed out perhaps as a symbol of sacrifice. On the other side there is the head of a bison bellowing with its tongue out as in the autumnal mating rut, four plant branches of perhaps conifer a whithering plant and three nut like seeds all indicating a composition of autumn.

Thus Marshak thinks from a study of Cro-Magnon Art and artefacts for a period of more than a decade and spanning practically the whole of Europe where ever these sites are located that the Cro-Magnon art means more than meets the eye. He thinks that these people had a sense of the passage of time and the seasonal variations and perhaps used the phases of moon to keep a count of this passage of time. They also had a sense of the differentiation between the sexes. He also thinks these people had different words to denote these differences. Thus Marshak tries to give a new interpretation to the Ice Age Man's art and artefacts and tries to peer into the working of their minds.

All animals can not survive in all ecological systems. Certain species are adapted to survive in certain ecological niches. However the emerging Homo was an omnivore and hunter. He did not depend on any particular animal for survival. He killed whatever animal he could and ate it. The animals however could not eat anything and survive. Some were browsers of leaves and some were grazers of grass. Some like temperate climates and some like cold climates. So they have a limited range and according to the seasons they have to migrate from one place to another more or less in a regular pattern. The hunter Homo also migrated with them. The animals also became scarcer or became extinct if the external environment shifted radically. This forced the Homo to migrate to other areas of the Earth where he could find better opportunities to hunt and gather edible plant food. It is this process which forced the Homo out of Africa and migrate to South East Asia, Northern China, North Africa and Europe some one million years ago. In the process he had to evolve also and improve his skills and techniques as well.

Thus the process of human evolution progressed from the Homos of Africa to the Homo Erectus of China and Java to the Neandertals of Europe and finally some 35,000 years ago to the Cro-Magnons of Europe. This does not mean that the other areas of Asia and Africa did not have any people. They certainly did. Only their remains did not survive the ravages of time. The archeologists and anthropologists being mostly Europeans, for a hundred years between roughly 1850 to 1950, they concentrated first on their home territories. Moreover the ice ages during the last 6,00,000 years peculiarly favoured France where the roaming herds of tundra were abundant. Here living in the caves and rock shelters of the limestone cliffs the Neandertals and Cro-Magnons evolved and developed their cultures.

It is not only Cro-Magnons who had an awareness of the animals and the ecology on which they depended for their survival. It is the emerging Homo from nearly one million years ago who was slowly having this consciousness as well as self consciousness about himself as well as the animals and plants and grasses on which the animals depend and which show variations with changing seasons and passage of time. Only we do not have any recognisable art or artefacts surviving before the Neandertal times of 1,00,000 years ago. That does not mean earlier Homos did not have any awareness of the nature of the animals they hunted or the grasses and leaves on which they depended or the seasonal variations which effect the grasses and leaf bearing plants or the water holes, rivers and lakes from which the animals as well as the Homos have to find drink. This awareness and its development and flowering is at the root of human evolution, more so of his brain capacity which we will deal with in the next chapter.

Between 10,000 and 8,000 years B.C the northern ice sheet retreated and melted down releasing vast quantities of water as well as land in Northern Europe and Northern Russia, Canada and north America. This vast quantity of water raised sea levels as much as by 300 feet and inundated much of the shore line of continents, which was exposed previously. This inundation separated England and Scandinavia, Russia and America at the Bering Straits. Indonesia was separated from mainland Asia and Australia from Tasmania and New Guinea. However man reached American continent and Australia much earlier than this separation.

Many lakes and rivers were exposed and their levels raised. All these waters as well as the shallow sea waters of continental shelves which were inundated were full of fish. We find for the first time now that dugout boats were made as well as many types of fish catching hooks and spears. Fish formed an important part of the diet of these people. The retreat of the tundra allowed the advance of hardwood forests northwards. Others like the mammoth died out completely, on the other hand plant food was abundant. The forest living animals were smaller and difficult to catch. Hence they had to device new techniques and implements to catch their prey.

The typical mesolithic tool was a microlith. It is a small sharp blade ground and polished and mass produced out of many types of materials not necessarily of flint. Then it could be fitted to any variety of weapon or tool heads as the business end of the weapon or tool. The most common mesolithic weapon is bow and arrow fitted with a microlith point bound by resin. They are also used to make composite weapons fitting the microliths to handles of bone or wood, thus making sickles to cut and gather the wild grasses that bore edible grains. Tools became more specialized to meet the varied needs of various environments and seasons. This enabled the mesolithic hunter gatherers to venture more farther north than their predecessors and make new lands habitable. They depended more on cultural innovation than on physical evolution. They were reduced in their body bulk, the size and strength of the teeth and jaw and its forward protrusion reduced. Their brain capacity increased, particularly in the brow portion. This produced the present look to the Homo Sapiens.

12

THE ROLE OF BRAIN IN HUMAN EVOLUTION

Of all the organisms on this Planet Earth man is undoubtedly the most dominant organism. Only bacteria and viruses were able to defy him until the 20th century. In the 20th century with the development of antibiotics and the advent of wide-spread immunisation man was able to control bacteria and viruses also. The other life forms like carnivores are no match for him. How this was achieved. He certainly did not have great bulk or limb power or claws or canines. His observational capacities of touch, taste, smell, hearing and vision are just about on par with those of monkeys and apes. Yet we are able to keep them in Zoos and conduct experiments with them in laboratories. What makes the difference. The difference lies in the way we could make use of the information we gather with our sense organs. Even though the apes and ourselves are equipped with the same kind of sense organs with the same range of development we could use the information gathered about our external world in a far superior and better manner than the apes and monkeys can do. This is due to the development of our brain particularly the cerebral cortex, that pinkish grey matter which is the much folded top layer of the brain in humans.

First we will take a look at the development of brain in lower organisms and its progress through the vertebrates, reptiles, mammals and primates to hominids. However, we are here primarily interested in the development of brain in hominids. In hominids only the cerebral cortex reached its maximum development which made humans unique among all living organisms.

The functions of all living organisms include the fulfilment of three biological necessities. They are to obtain food from their surroundings, protect themselves from predators and seek and attract mates to reproduce their kind. They seek instinctively or in the case of humans consciously to the fulfilment of these biological necessities. For that the organisms should first obtain information about their surroundings. This information should be processed and appropriate reactions should be taken at the right time and place to be successful. Any mistake will result in failure. A proper stimulus response mechanism is essential.

To obtain information from the surroundings the organisms are provided with sense organs. The principal senses are touch, sound, taste, smell and vision. Some organisms have one or two senses developed better than the other senses to suit their particular needs. Fish have a developed sense of taste and touch. Most insects depend upon smell and taste. Many animals like rabbits, bats and some foxes depend on a keen sense of hearing. The principal sense on which mammals depend is smell. They have extra large muzzles provided with a network of nerves sensitive to smell. Some of them secrete special scents and also use urine to identify themselves and their kind.

Vision is one of the basic senses of animal kingdom. However not all animals see the same way. They see the same thing in different ways. The reason is that many animals can not distinguish colours. Besides many animals have their eyes at the side of their heads. This gives a wider field of vision but poor details. Only certain higher animals like the primates have their eyes in the front of the skull in the same plane. This gives them a three dimensional vision. Besides higher primates like monkeys and apes also have colour vision. Man however has all these senses developed to appropriate levels. His principal sense is vision.

The sense organs themselves do not see, hear, smell, taste or feel the pressure etc. There are nerve cells in these organs. They feel the respective stimulations from the outside world. These stimulations generate minute electro chemical impulses which travel through the nervous system to the brain of the organisms. The appropriate areas in the brain process these stimulations and coordinate them with stimulations from other sense organs and send appropriate orders again through the central nervous system to the muscles for reaction. This is the basic principle of stimulus reaction mechanism. But many lower organisms have no brains at all. A few nerve cells, formed into a knot act as brain. We are here concerned with the brains of vertebrates and mammals only. Our primary interest here is to study the development of hominid brain which lead onwards to homo sapiens *i.e.*, ourselves. The starting point is mammalian brain.

In the vertebrates and more so mammals the brain is located in the protective cranium of the skull. The skull is joined to the tail by the vertebral column and in between joining the thoracic and pelvic girdles which are supported by legs. The nerves from various organs are formed into a bundle and pass through a cavity in the vertebral column to the brain. This is the central switch board and the nerves are the wiring which takes messages to the brain and in turn bring orders to the various organs. Without this elaborate system the body cannot function. The realization that the brain controls the activities of each and every part of the body and its organs came during the second half of 19th century. Since then elaborate efforts were made to understand which part of the brain controls what activity and this process is still going on.

These actions are both involuntary and voluntary. In other animals these activities are mostly instinctive. In humans the voluntary activities altogether acquire a new dimension. They are mediated upon and conscious. The brain development in the hominids brought about this change. Instinctive action is relegated and mediated action is promoted. This also reduced the necessity for physical evolution and developed those brain centres which gradually made the hominids independent of physical evolution. These centers are now understood to be located in the cerebral cortex, that pinkish grey top layer of cerebrum.

The nervous chord which passes through a hole in the vertebral column enters the brain through a hole at the base of the skull known as the foramen magnum. We have seen in earlier chapters the position of this foramen magnum is an important indicator whether the hominid was bipedal or struggling to be a bipedal. The nerve chord ends in a bulging knot in the centre of the brain known as the brain stem. It controls such vital involuntary activities as the breathing, blood pressure, swallowing, heart rate, alertness, wakefulness and sleep. The hypothalamus is situated above the brain stem and it controls the constant body temperature, food and water requirements and sex and responses to threat. It also controls the activity of the pituitary gland which hangs from the hypothalamus. It activates the pituitary gland to release the hormones that are necessary for growth of the body through their control of the metabolism. They also control sexual activity, pregnancy and reproductive cycles in females as well as their lactation by activating the mammary glands. The hypothalamus also regulates body heat which is so essential for mammals to facilitate all those critical chemical reactions to maintain their body metabolism and other functions. When the body is too hot it activates the sweat glands to cool the body. If the body is too cold it narrows down the capillary blood vessels near the surface of the skin to prevent heat loss.

The thalamus is a structure in the mid brain next to the hypothalamus. Its main function is to act as a relay station for the messages received from various organs and send them to the appropriate brain centres which process them. The cerebellum is a structure at the back and base of the skull. It coordinates all our movements and keep us in balance. The pons connects it with cerebrum and cortex. The limbic system which is also part of the mid brain control emotions and some memory. All the systems described so far are basic and primitive in the sense they evolved very early in the life of vertebrates and mammals. What is unique to mammals and primates is cerebrum and cerebral cortex.

The cerebrum is the dome shaped structure above the mid brain and it is prominent only in higher primates and hominids. In lower mammals it is rudimentary and in higher mammals like cats and dogs it appears as a swollen corrugated layer. The cerebrum is 40% in brain weight. The top layer of cerebrum is the cerebral cortex which contains some 75% of the 100 billion neurons or brain cells. This is the centre which gives us conceptual thought, language, memory, plan, purpose, a sense of values and other complicated abilities which distinguish us as humans. The cerebrum is divided into two hemispheres which are connected by corpus collosum which allows us to function as a single individual instead of a split personality. The two hemispheres normally divide the work of the brain. The left brain controls the language ability as well as our analytical logical and reasoning activities. The right side usually control our integrating abilities, our emotions, facial memories and our musical abilities.

In the cerebral cortex there are two strips of brain area extending from ear to ear over the top of the brain. The back strip receives sensory signals from the parts of the body. The front strip is the motor cortex which activates the muscles and proper responses to the sensations are sent back. The lobe at the rear of the brain receives and controls the sensations of vision. Auditory cortex is located on either side of the brain above ear. The taste and smell cortices are located nearby to the auditory cortex.

Language depends on two areas of the left hemisphere. Broca's area named after the French surgeon is located on the left frontal cortex. It controls and integrates the muscles involved in speech, those of tongue, lips, jaw and throat. Another region the Wernike's region is located on the left temporal cortex and is involved in understanding the language. The development of these regions in the skull together with other indications in the thorax and vertebral bone named T7 give clues to the scientists whether the hominid developed any ability for communication using language.

The cerebral cortex is about a millimeter thick and 1.5 sq. meters in area if spread flat. It is much folded and convoluted in the brain cavity. The convolutions and folds and furrows are similar in all men. There are 28 bones in the skull of which 8 form the vault of the brain, their joints interlocking in a jig saw fashion. The lobe of the brain which is nearest the skull plate is called by the name of that plate. Thus the occipital plate is at the rear of the skull and the part of the brain at the back is known as the occipital lobe and contains the centres of vision. The two skull plates at the top of the skull on either side of the crest are known as the parietal plates and the brain lobes nearest to them are called the parietal lobes. They keep control over the body parts and their movements. The plates above the ear and below the parietal lobes on either side of the skull are known as temporal plates and the brain lobes near them are known as the temporal lobes. They control hearing. The skull plate of the brow covers the frontal lobe. It is believed that herein lies many of man's unique features like conceptual thought, foresight, ethics and spirituality etc. However still many parts of the brain's cortex are unmapped. We are unique in the living organisms because we have evolved this cortex.

We have to remember here that absolute weight of the brain is not the criteria for the advanced nature of the brain. It is the ratio of brain to body weight that is the criteria. Also the brain of elephants and whales are larger than man's brain. The brain body ratio is 1 : 50 in man, 1 : 250 in elephants and 1 : 1000 in whales. One gram of elephant brain has to work for 250 gr. of body and 1 gr. of whale brain has to work for 1000 gr. of whale's body which make them inferior to human brain where 1 gr. of human brain has to work for only 50 gr. of body.

We will now consider the development of brain in mammals, primates and hominids. The chief characteristic of the mammals is to give birth to the young and nurse them with milk. This system of improved reproduction requires many corresponding changes in the bodies as well as brains of the mammals. The hypothalamus is the principal part of the brain which triggers the release of the hormones which governs the ovulation cycle, sex life, pregnancy and child birth as well as the lactation of the mother. The capacity for body heat regulation at a constant temperature is very essential for the complex activities of the cerebral cortex as well as memory which plays such a vital role in the life of mammals. Mammals could also express emotions like love and anger. The limbic system which is part of the mid brain and separates cortical hemispheres above it from the mid brain is responsible for emotions. It is also responsible for some memory capacity. The reptiles lack this limbic system and consequently do not have emotions and poorly developed memory.

The primates which adapted themselves to life in trees necessitated corresponding changes and developments in the brain. One of the developments is the movements of the eyes to the front from the sides of the skull. This gave a face to the primates. The front eyes in the same plane produced two images which overlapped on each other in the brain, thus giving three dimensional vision to the higher primates. This helps estimate the distance of the objects correctly and help the primates in the trees. The frontal eyes also reduced the importance of smell and consequently the size of the snout reduced. Also the size of the brain cavity increased. In the brain itself the area devoted to smell reduced and the area devoted to vision increased.

Another important feature of primate life is the coordination of brain functions controlling the eyes with those areas of the brain which control the hand and leg movements. This is important for the primates' survival as they jump along the branches of trees and vines high above the ground. This is achieved in the area to the rear of the frontal lobe. The organization of the brain matter is different and much improved in primates than in other mammals. The monkeys for the first time diverted the evolutionary changes from body parts to brain. This is a very important contribution of monkeys and apes to the further evolution of hominids.

Among the apes the gorillas and chimpanzees are closer to us than the other two Asian apes of orang-utans and gibbons. The gorillas separated some 9 million years ago and the chimpanzees separated some 6 million years ago from the stalk that finally led onwards to hominids and homos and man. Africa is the home of gorillas and chimpanzees. So the early hominids also evolved in Africa particularly in East Africa. Conditions in East Africa changed some 5 or 6 million years ago due to rifting and elevation of the eastern portion of the continent. Cool and dry conditions prevailed due to the formation of ice at Antarctic and Arctic. The forests have broken up into patches of forest, bush country, savanna and arid zones. Some of these ape ancestors of early hominids had to develop erect posture to survive in the changed environment. There are no good hominid fossil skulls of this ancestry to estimate their size or shape.*

The next oldest hominid fossils are those found by Donald C. Johanson in the Hadar region of Ethiopia now made famous by the name "Lucy". They are estimated at 3.2 to 3.5 million years old. Other specimens of older ancestry came from Leotoli region of Tanzania discovered by the Leakey family of Richard and his wife Meave whose age was estimated at 4.1 million years. Tim White of U.S.A. discovered in Ethiopia some hominid fossils which were estimated to have even older age.

The skull capacity of these hominids were measured and Tobias concludes that their skull capacity and brain size was not greater than chimpanzees. These hominids were named A. Afarensis after the Afar region in Ethiopia. The mean brain size of A. Afarensis is 413 C.C. whereas that of chimpanzee is 380 C.C. and Gorilla is 500 C.C. Allowing for the bulk of the gorilla which is larger than chimpanzee as well as A. afarensis, afarensis brain is not more developed than that of these two African apes, concludes Tobias.

Thus there were what are now grouped together as Australopithecus Afarensis group of hominids extending from Ethiopia to Tanzania to South Africa living between 4 to 2 million years ago. There may be more species of hominids living in Africa evolved as a result of the environmental changes and adaptive radiation to suit such changes. Under such circumstances only some species remain to evolve further and many others die out. The Paranthropus and Zinjanthropus which were small brained but with large jaws and dentition and primarily adapted to a vegetarian diet died out. The A. Africanus variety which had a little more improvement in brain capacity than A. Afarensis gradually adapted to a mixed diet and seemed to have evolved further. A. Africanus brain capacity is 430 C.C. only a little more than that of modern chimpanzees. Considering their smaller size also some scientists think that their brain capacity is 30 to 40 per cent more than the chimpanzee brain. They might have walked

---

* Phillip V. Tobias: *The Brain of the First Hominids*, p. 62, *A Symposium of the Foundation Fyssen*: edited by Jean Pierre Changeux and Jean Chavaillon: Clarendon Press, Oxford 1996.

with a shuffling gait and still retained the ability to climb trees. They might be shifting from a vegetarian diet to a mixed diet of termites, grubs, bird's eggs and small game like antelopes, bush pigs and small monkeys like chimpanzees do even today. Tool use is not definitely established. Communication and group living and co-operation with each other also may not have exceeded much over those of modern apes. The hominids living and evolving between 4 million and 2 million are closer to apes than to man.

Hominids with larger brains evolved between 2 and 1.5 million years. They are now known as Homo Habilis. The first fragments of H. Habilis were found by Jonathan Leakey, eldest son of Louis and Mary Leakey, in a branch of the Olduvai Gorge in 1960. Latter several other fossil fragments were found. Its brain capacity was 640 cc compared to 430 cc of A. Africanus. Homo habilis fossil fragments accumulated over the years and there is much variation in its size as well as brain capacity. A particularly fine specimen of H. Habilis skull KNMER-1470 discovered by Richard Leakey, second son of Louis and Mary, in 1972 on the eastern shore of Lake Turkana exhibited a skull capacity of 770 cc. For this great variability H. Habilis is considered as intermediary variety of hominid evolving quite fast between the A. Africanus and Homo Erectus which came latter. What triggered his evolution then. Scientists are of the opinion that another cold and dry spell engulfed the world around 2.5 million years ago due to the formation of more ice at Antarctic and Arctic.

Richard Leakey and Alan Walker of U.S.A. unearthed the fossil skeleton of what is now known as "Turkana Boy" in 1984 on the western shore of Lake Turkana in Kenya. It was almost a complete skeleton without feet and hands and estimated to belong to a boy of 12 years age. The skeleton was quite tall and the cranial capacity was estimated at 900 CC and the age at 1.6 million years. This is named as Homo Erectus, the same species that were found earlier in Java and Peking. It had a larger brain capacity and consequently better equipped to live in the dry and arid conditions of East Africa living mostly by hunting small game. Meat eating contributed greatly in the development of brain because brain tissue requires nearly 5 times the energy of body tissue and the energy rich fat laden meat, marrow and brains of the animals provided the source of this concentrated food.

Then between 1 million and 500 thousand years we find this Homo Erectus migrating to as diverse places as North Africa, Europe, China and the far East. In the process we find this H. Erectus developed larger brain, better tools, hunting and eating larger game. Tobias gives the mean brain capacity of these Homo Erectus at 930 cc with the Peking specimens showing as great as 1040 cc. The use of fire is the main cultural advancement we find with the Peking variety of H. Erectus. The tools were also much refined to produce a core, chipping away allround the stone. There is evidence of cooperative hunts of big game by fire drives etc. in Europe.

All in all the Homo Erectus of East Africa ranged far and wide in the habitable world of his day. This is possible due to his evolving brain which allowed him to hunt and gather a more diverse diet than the restricted diet of other species. He was an omnivore enjoying a variety of foods, making better tools. What is more he learned the advantages of cooperation with each other. He invented the use of fire, which allowed him to roast meat, live in cold climates, ward off predators. His 900 to 1100 brain capacity allowed him to do all these advanced tasks. No need to say that he had a better repertoire of communication capabilities with each other though we can not say it is true speech.

Between 500 and 150 thousand years ago there are not many fossil specimens available. We have only some fragments of skulls from Steinheim and Heidelburg in Germany and Swanscombe in England and some other places. They are now called Archaic Homo Sapiens. They are estimated to have developed a better speaking ability than Homo Erectus.

Homo habilis is now associated with chipped stone tools we call Oldowan industry, pebbles chipped at one end to make jagged edge, and the chipped away fragments used as scrapers. Tool making implies some degree of conceptualisation. We could also say that homo habilis exhibits a degree of development in the Broca's area. Ralph Halloway a specialist in Hominid brain casts from U.S.A. identified the development of Broca's area in the skull cast of E.R. 1470 skull. It is an indicator that the H. Habilis was able to communicate with each other using at least a degree of sound symbols to objects and feelings beyond the simple grunts and groans of chimpanzees. Any association of sound symbols with objects and feelings involves abstraction and conceptualisation. We could say that the increased brain size of homo habilis set it on the road to conceptual thought which is the most characteristic feature of Homo Sapiens.

We find a human form who is named Neandertal man by about 125 thousand years ago in the valleys of Western Europe. He was rather short and bulky, with a receding forehead bulky brow ridges, heavy and protruding jaw, powerful torso, arms and legs, his brain capacity was about 1350 C.C. The skull itself was flatter at the top, receding in the fore head and wide and long at the base, with this increased brain matter he developed a remarkable culture. There was tremendous improvement in his artefacts, his tools, his dwelling places, and his burial practices of the dead, none of which were found earlier. He gave a careful and reverential burial to his dead, daubing the dead man with red ochre and placing flowers around him. This indicates development of consciousness and self-awareness which are the hallmarks of Homo Sapiens. His treatment of the injured and the sick also show a degree of empathy between members of the group at a higher level not found in H. Erectus. All this indicates an increasing level of social life and communication, self awareness, consciousness and compassion. These qualities are not available in such an emphatic way in H. Erectus or H. Habilis. They are intermediary to apes and Homo Sapiens, though closer to H. Sapiens. With Neandertals and their 1350 C.C. brain power we can say that hominids have entered the phase Homo Sapiens the last phase on the long evolutionary story of mankind.

The Neandertal man was gradually replaced by another kind of man from Eastern Europe, Near East and Africa. Whether this has taken place as a result of extinction or replacement of the Neandertal man is not known for sure. There were no Neandertal fossils after about 35 thousand years ago. The Cro-Magnon man who replaced the Neandertal man was physically akin to modern man. He had a broad and high forehead, the skull was high domed and less wide at the base. The occipital bun at the back of the skull which was so characteristic of Neandertals was absent and there is a prominent chin present. The skull capacity on the average is 1350 C.C. We the modern men, Homo Sapiens are his heirs. We know what we did with our large brain during the last 35,000 years. The Cro-Magnon man was a superb artist, cultural and technological innovator. He was on the road of replacing the necessity of laborious physiological evolution with much faster cultural innovations. He further developed social organization, specialization of work, division of labour and perhaps exchange of goods. Communication through sound symbols was developed which ultimately became language and true speech. His experiences became cumulative which could be passed on to succeeding generations thus building up a wealth of information, tradition and mythology which supplied the social and behaviour codes and moral and ethical norms were developed. His consciousness developed which led to an intense awareness about himself and his surroundings and stirred in him a longing to search for a meaning for all that he sees and experiences and what his place and role in it.

The Cro-Magnon man spread throughout the breadth and width of the habitable earth and his spread through varying physical surroundings and environments produced racial and cultural differentiation but not separate species. Cro-Magnon man remained a single species in spite of several races. All these races share a common genetic endowment. Cro-Magnon man's 1350 C.C. brain capacity gave him a plastic behaviour which could adapt. Thus he replaced physical evolution with cultural adaptation.

13

HOMO SAPIENS AND HIS UNIQUE EVOLUTION

Man has become human in a social environment. He acquired his distinctive human qualities while living in cooperation with others of his kind. Social life is the setting in which he evolved use of tools, conceptual thought and language, division of labour, family life, self consciousness and social order and cumulative tradition.

His unique biological features have evolved earlier under norms of biological evolution *i.e.*, natural selection, in a limited semi social environment varying between semi ape and semi man. This physiological evolution is slow as all biological evolutions are and took place mostly during the past 3.5 million years. Between 3.5 to 2.5 million years, the hominids became fully erect and hands became free to manipulate objects. This is the stage we now call Australopithecus Afarensis which includes Ethiopian, East African and the South African varieties. The Paranthropus of South Africa and the zinjanthropus of Tanzania died out and eventually the smaller and more brainier varieties evolved further. Australopithecines had a brain capacity of 430 c.c. and they were not associated with any cultural artefacts like tools.

The Homo Habilis skull labelled E.R. 1470 and discovered by Richard Leakey had a brain capacity of 770 c.c. and aged at a little less than 2 million years. It is associated with crude chopping tools. The chopping tool requires a conscious operation on it, often performing it on the mental image. That is possible only with a larger brain. Our nearest animal the chimpanzee can not retain images for more than a few minutes with its less than 400 c.c. of brain capacity. Almost double that size is required to conceptualize objects and perform mental operations on them. This is the beginning of conceptual thought and culture.

By 2.5 million years ago the environmental changes produced more ice at the poles and rifting in East Africa. This resulted in more cooling and drying up of the region. The antelopes living at this period in East Africa between 2.5 and 2.0 million years show remarkable evolutionary changes. So do the hominids. Their brain capacity doubled from 430 c.c to 900 c.c. by 1.5 million years ago, the stage where Homo Erectus had evolved. This is rapid change, doubling of the brain capacity in one million years. One million years may be eons and eons of time by human standards but by evolutionary standards it is only a fraction. During the next million years between 1.5 million to 0.5 million years when we have the fossils from Zhoukodian caves in China available, the brain capacity increased from 900 c.c. to 1040 c.c. This is another 15% increment.

The brain development between 500 thousand to 100 thousand by which time we have Neandertal type of Hominid evolved, is from 1040 c.c. to 1350 c.c. or about 20% increment. This is in the range of Homo Sapiens but the shape of Neandertal skull is different and consequently the arrangement of brain matter inside could also be different. That is perhaps one reason he became extinct unable to compete with the latter day Cro-Magnons who are akin to us.

What is the basis of this fast development between 2.5 to 1.5 million years, when the brain capacity doubled. There has to be some evolutionary opportunity available to the hominids by 2.5 million years ago. In the past evolutionary eras such fast evolutionary development happened when a new opportunity or a new food source was exploited by an animal which has fortuitously found itself in a position to exploit it.

The rapidly changing environment by about 2.5 million years forced the hominid to direct its attention to the environment itself. Its changing seasons, abundance or lack of food with changing seasons, migrating animals with grass and herbage changing locations with changes in atmosphere forced the hominid to focus its attention on the changes in nature itself. These changes are not random. They are cyclical, cause is followed by effect. There is a spatio-temporal sequence present in these changes. Those who mastered this cause and effect relationship had a great advantage. Let us call this deductive logic or causation present in nature and the evolving hominid had mastered this process and greatly benefited by it. It helped him evolve his brain capacity from 430 c.c. to almost 900 c.c. by 1.5 million years, or from Australopithecus to Homo Erectus.

His ability to stand erect and use hands to manipulate objects, to use sticks and long bones to dig roots and tubers were greatly enhanced and resulted finally in making of crude chopper tools with stone. Such crude tools began to appear by 2.4 million years and by 2.0 million years they were definitely established with the Homo Habilis and his roughly 770 c.c. brain capacity. This tool making is the beginning of culture and conceptual thought. Homo Habilis is an intermediary stage between Australopithecines and Homo Erectus which made its appearance in East Africa by 1.6 million years. The tools of Oldowan culture which made their appearance by 2.4 million years were firmly established by 2.0 million years. These tools are water worn pebbles with a few flakes struck off one end. This could be used to smash bones for marrow. The struck off chips could be used as scrapers to cut pieces of meat from carcasses. These hominids between 2.5 and 1.5 million years in East Africa mostly depended on scavenging the kills of big cats and their own hunting is much limited to small game. However they are gradually shifting to a diet of meat eating. The sites at which tools are found indicate they did not venture far off from water sources.

The Homo Erectus which evolved by 1.6 million years, on the other hand started migrating out of Africa. His larger brain as well as his larger body size and quite erect as well as tall and lithe frame allowed his migration. Meat eating helped this migration greatly. Not only he migrated but his prey animals as well as some of the carnivores also migrated. Such evidence of migration was found in Europe and Asia after 1 million years only. Better tools known as hand axes associated with Homo Erectus began to appear in Europe.

Stone tools known as bifaces began to appear in Asia and China. These are better chopping and cutting tools than the Oldowan tools. A slain animal can be skinned and chopped with them easily.

The Homo Erectus emerged by 1.6 million years ago and the skeleton of a 12 years old boy was unearthed by Richard Leakey and Alan Walker in 1984. It was the earliest specimen. Homo Erectus migrated out of Africa to Europe, China and South East Asia. His way of life was hunting and gathering. He became meat eater which facilitated his migration out of Africa. He made finer tools, regularly hunted medium sized animals like deer and above all mastered the use of fire. His brain capacity is between 900 c.c. to 1100 c.c. He is leaving behind his ape qualities and acquiring human qualities. He had developed a home base where food is brought and shared. This required several cultural developments and innovations.

Discoveries* in the 1990s in Georgia a former Soviet Republic however revealed the migration of hominids into Europe started even much earlier than 1 million years ago. The skull found at Dmanisi in Georgia is of smaller proportions than the Homo Erectus skull. Its age is estimated at 1.75 million years. Unless, other body bone fossils are also found, it would be difficult for anthropologists to agree that a creature like Homo Habilis with shorter legs and longer dangling hands and a brain capacity less than 700 c.c and with tools no better than Oldowan choppers and scrapers ventured out of Africa at 1.75 million years ago. But the skull is undisputably much older and smaller than that of Homo Erectus skull.

Homo Erectus is quite a family fellow. He developed hunter gatherer way of life. He maintained a home base where he could return with the kills. The females gathered fruits, berries, nuts and tubers. Thus a division of labour was established. The infants of H. Erectus also required longer period of sustenance and care taking by the mother than the earlier varieties of hominids. H. Erectus growth pattern is shifting towards humans. He is taking more years to mature. This home base resulting in a hunter gatherer way of life much like the bush people of the Kalahari desert of Africa now

* Rick Gore: *National Geographic*, August 2002.

resulted in a number of social innovations. Food has to be brought home and shared now. How much goes to whom and the behaviour pattern and code of conduct involved in it is quite complicated. Sharing and gathering and hunting also requires territoriality we watch among apes as well as other animals. This often involves complex relations between several groups. When food is scarce this may even result in head bashing and eating the slain in a ritual. We find several H. Erectus fossils with skulls bashed. We also find evidence of cannibalism. So the way of hunter gatherer life evolved by H. Erectus is not so innocuous as it looks. It is very complicated. It helped evolve his social, family and cultural life. It continued to be the basis of cultural as well as physical evolution until we come to the neolithic societies by about 10 to 12 thousand years ago. Development of language and consciousness depended on a social and co-operative way of life necessitated by hunter gatherer societies.

We observe that the H. Erectus of Zhoukodian caves near Beijing in China developed use and control of fire. We observe the H. Erectus of Torralba and Ambrona in Spain also used fire drives to drive elephants into bogs to mire them and kill them. This control of fire gave them roasted meat which is easier to slice, eat and digest. It gave him warmth and allowed him to migrate into colder regions. It gave him protection from predators. We find this control of fire around 500 thousand years ago. We also note that starting around 600 thousand years much of north America and Europe were covered by glaciers resulting in a cooling down of temperatures world wide. This naturally forced the hominid to find ways and means to combat it. By this time his brain capacity also developed to about 1000 c.c.

This control of fire was made possible because man was mastering the deductive logic available in the repetitive events of nature. This was made possible by his long memory associated with a large brain.

Language ability however was the most characteristic and defining feature of H. Erectus and subsequent hominids. Language ability is controlled by lips, tongue and the vocal chords in the throat. They are soft tissue that do not leave fossils. However there are two other features in the brain that show themselves in fossils. The Broca's area located as a small bump on the left frontal lobe above the ear is identified as the brain area which controls the muscles of tongue, lips and throat associated with language. It is estimated by experts that H. Erectus shows some development of this Broca's area. Another area which is on the left temporal lobe behind the ear is known as Wernikie's area. This is known as involved in understanding of language. A third area in the skull which has a role in language ability is at the base of the skull. The vocal chords are located quite high up in the throat of apes whereas in humans this is located downwards towards the middle of the throat. Now surgeons and anatomists found a connection between the location of the vocal chords in the throat and the bone at the base of the skull by the degree of downward bending this basicranial bone shows. This bone is bent downwards more in humans than in apes. All these indicators point out that the H. Erectus had some degree of vocalization available to it, though we can not say definitely it is true speech.

The earliest speech could have been about giving sound symbols to some objects and activities *i.e.*, developing some nouns and verbs. The objects they are occupied daily in their lives are the animals they hunt and the fruits and berries they gather. These could be associated with some sound symbols without much difficulty provided the hominid had large enough brain to abstract and conceptualise the object or some of its qualities. Next he could give sound symbols to verbs like the animal is running or walking or standing. This also leads to adjectives and adverbs like big and small fast and slow. If a group of animals are under observation that leads to a necessity to develop prepositions and conjunctions like before and together and after. This is of course an imaginary situation and it is not far difficult to imagine many such situations where word symbols could have developed.

For words to develop concepts should develop first. That is the image of an object should be abstracted and retained in our memory long enough to be able to perform some operations on it or to play with it so to say. This requires neuronal development in our brain to a level where we could retain the memory image long enough. Not only retaining an image but its recall whenever required is also necessary. We also know that Australopithecines by about 2.5 million years ago did not have a brain capacity or body size much larger than the chimpanzees. So he could not have word capacity.

The earliest evidence we have about abstraction and conceptualisation was 2.4 million years ago when we have some glimmerings of stone tools. Making a stone tool requires some degree of abstraction and conceptualisation. Some mental operations should be performed on the stone before a tool is made. He should be able to visualise the tool before he makes it. He should carry the stones from their natural settings to where they are needed, to places often where the stones are not found in nature. Any and every stone is not suitable for making stone tools. The earliest stone tools are water worn pebbles gathered from river and stream beds. So, without a set purpose in mind the hominids of 2.4 million years ago would not bother to gather these stones and rough up one edge. This is the beginning of conceptualisation.

By 2 million years ago we have definite stone tools, made to a standard pattern now called the Oldowan industry. These tools are associated wilth Homo Habilis an intermediary hominid between Australopithecus and Homo Erectus. The best known specimen of this hominid is E.R. 1470 whose brain capacity is estimated at 770 c.c. The next increment of brain capacity is in Homo Erectus which we have by 1.6 million years ago and its brain capactiy is estimated at 900 c.c. So something is happening very fast between 2.5 million to 2 million years and 2 millions to 1.5 million years ago. This is an increment in brain capacity, together with an increment in body size as well as a more erect posture and gait with a lithe and tall frame. Homo Erectus has arrived. He is fully well set on the road to humanity.

This increment in brain size from 430 c.c. by 2.5 million years to 900 c.c. by 1.5 million years when H. Erectus evolved to 1100 c.c. by 500 thousand years in China variety of H. Erectus and 1350 c.c by 100 thousand years in Neandertals and about 1400 c.c. by 40 thousand years in Cro-Magnons is fast change in the beginning to some slowing down in the last phase. There has to be some advantage in evolving larger brains or more neurons in the brain. Most of this evolution is in the cerebral cortex and cerebrum. Some outside factor has to exert evolutionary pressure and the subsequent selection of this neuronal increment. A direct result of this neuronal increment is increment in memory and its recall.

Memory is possible with a repeated perception. The perception has to create a neuronal pattern in the brain. Repeated perception creates a template of neurons that is stored as memory. What repeated perceptions are there then available to the hominids by 2.5 million years. The African landscape and environmental conditions are fast changing at that time forcing the hominids to direct their attention to the environment. This gives them memories about external conditions which they are forced to remember longer and recall when necessary. This is possible when there is repetition in natural events and there is advantage and evolutionary selection in remembering such events and their sequence. Any hominid which had a little more number of neurons in the brain remembers a little longer. Thus increment in brain matter which had a selection value depends or takes advantage of the causal connections and spatio-temporal sequences present in nature. Such genes quickly spread in the inter breeding deme and form the basis for further selection pressure. Thus, increment in brain matter gave the hominids tools first and hunting gathering way of life, meat eating, co-operative and social living, mastery of fire, language and awareness and consciousness and a cumulative tradition successively.

Then why the growth of brain stopped with modern man at a mean racial level of 1350 c.c. There are two reasons. There is a mechanical reason which does not permit a larger skull to pass through the birth canal of humans. This forces the skull to be more pliable and flexible and contain lesser matter at birth which necessitates longer years of development and postnatal care and dependence. The second and more important reason is that Homo Sapiens with his large brain passed the limitations of evolutionary mechanism and replaced it with cultural adaptations to achieve his ends and goals. There is no need for further enlargement of neuronal matter in the brain. He depended on culture for further evolution. His 1350 c.c. brain was good enough for producing culture.

In the absence of clear evidence let us say that H. Erectus started creating words for objects and feelings and some verbs, adjectives and adverbs, prepositions and conjunctions starting between 1.5 million years and 1.0 million years. This increased communication ability greatly facilitated his migration out of Africa and into Europe and Asia. Thus he set himself on the road of cultural adaptation. This cultural adaptation is the basis of his further evolution which is not merely on the physiological plane but on an entirely different plane that of cultural plane.

This is the period of Archaic Homo Sapiens or transitory Homo Sapiens between Homo Erectus and Neandertals. Archaic Homo Sapiens fossils were found at many places starting from South Africa, East Africa, Mediterranean Africa to Europe and China. Of particular interest are two partial skulls one from Swanscombe, England and another from Steinheim, Germany and a mandible from Heidelberg Germany. By this time between 500 thousand and 250 thousand years ago the Archaic Homo Sapiens might have begun to assemble simple sentences with words already deveoped by his predecessors. By 250 thousand we have Neandertal type of fossils appearing and his brain capacity is 1350 c.c in good specimens which appeared much latter.

The increase in brain capacity developed starting from 1.5 million years ago when it was 900 c.c. in first H. Erectus specimens to 1100 c.c in China H. Erectus and to 1350 c.c in Neandertals by 100 thousand years and there is no appreciable difference between Neandertals and Cro-Magnons at all. This is due to cultural adaptation replacing the slow biological evolution as a means of making Homo Sapiens and modern humans who are designated by Homo Sapiens. Then we have to take a closer look at this cultural adaptation and its antecedents.

Culture is something unique to man. This depends on communication using words for objects, feelings, experiences etc., and the words when developed with grammar and syntax their communication extends to a veritable field of human experience. This is the basis of cultural life, sharing experience in a society. Without society there is no culture. Society is the medium through which culture operates. We can even say that society is that group which shares a common body of inter communicated experiences and the behaviour shaped on such experience. Human society is unique in the fact that it promotes further human evolution, particularly his cultural evolution. Cultural evolution takes place by awakening human potentialities for social life, self awareness and development of ethical, moral and spiritual life. By such awakening, larger and larger social groups are formed by an amalgamation of smaller social groups into larger social groups. This process is what has taken place in human societies during the last 100 thousand years or more and is still going on now struggling to weld the entire human population of the planet earth into one global society. World wide communication and internet facilities, world wide trade, exchange of goods and ideas is playing a great role in this process.

So what is central to such world wide amalgamation is symbols, or words language in short. Language needs the evolution of abstracting and conceptualising capacities. We have observed earlier that glimmerings of such capacities were evident by 2.4 million years ago when we have some very crude stone tools. There are definitely made Oldowan tools by 1.9 million years when we could say that these abstracting capacities have developed further. By 1 million years H. Erectus began to wander, made better tools, developed hunter gatherer way of life, developed a home base where food is brought and shared. This required complex social relations to be formed to evolve a cooperative social living. By this time he had many words for his simple needs of communication. He also had a brain capacity of nearly 1000 c.c. By 250 thousand years, he is estimated to have simple sentences. By 100 thousand years his brain capacity reached 1350 c.c and he ought to have developed true speech. All this is possible by the development of his abstracting and conceptualising capabilities and increment in brain capacity acting on each other in a feedback mechanism. Let us now have a look at this mechanism.

Conceptual thought is the basis for developments in the hominid evolution during the last 2.5 million years. Conceptual thought evolved taking advantage of the spatio-temporal causal connections in nature which are repetitive. Hominids have five important senses on which they depend for their information about outside world. They are taste, touch, smell, hearing and the most important of all vision. Perception is stimulus to one or more of our sense organs, concept is the memory image of that stimulus in brain. A single perception does not produce a concept. Repeated perceptions are required for concept formation. A single perception can not form the basis of a concept since perceptions often deceive us or misrepresent the source of the stimulus. So, for any proper concept to be acceptable should be repeatedly perceived not only by one person but by a number of normal healthy human beings. Otherwise even hallucinations would pass on as concepts.

Man, like all other organisms which took advantage of some factor present in nature to evolve themselves, also took advantage of the cyclical causality present in nature to evolve from ape man to full man. Tool use, conceptual thought, use of language, cumulative culture all presuppose a mastery of the causality, order present in nature which is repetitive. No other animal could do it because they lack the combination of antecedents required for it. In the words of *Julian Huxley, conceptual thought could not have evolved in any other animal but man.

Larger brain capacity gave longer memory. Longer memory remembered repetitive events and the association or bunching or sequence or one following the other of events. That enables him to anticipate the succeeding event when he observes the preceding event. When a lion or a tiger waits at a water hole, it anticipates the arrival of other game animals on which it could pounce for a kill. Man could do it in more situations which require longer memory and which have sequencing. This gives him a great power over all other animals, as well as nature, enabling him to survive much better than any other member of the animal world.

The development of emerging man's brain capacity gave him long memory and long memory remembered previous events and their repetitive character. Sun rises every day. Moon waxes and wanes every 29 days. The change of position

* Julian Huxley, *Man in the Modern World* - Mentor Books.

of the sun on the horizon gives rise to seasons. Seasonal changes are important for the variation in temperature and day length which are in turn responsible for summer or winter and abundance or scarcity of food. In the northern hemisphere with thick ice sheets covering large parts of it, the association of seasonal changes with Sun's position on the horizon and day length is all the more dramatic. It is there that important developments took place during the last 1.0 million years.

With a large brain and long memory man could not escape the repetitive nature and connecting links between various events. It is only one step away from deductive logic and man depended on it heavily in his evolution from ape man to full man.

Long memory, concept formation and imagination combining simple concepts and communication with other members of the group using language and building up cumulative tradition are the result of this mastery over the sequential events in nature. Other cultural innovations and adaptations followed, the chief among them being the mastery and control and use of fire. The development of more refined tool kits to suit various seasonal and ecological needs and developing and adapting dresses, homes and hearths followed. All these developments are made possible by mastering the deductive logic available in the repetitive events of nature, facilitated by his large brain which evolved concurrently with them.

These capabilities, large brain, long memory, conceptual thought, communication through language, cumulative tradition, cooperating and social living in large groups made man unique among all other organisms in nature. He was able to achieve this by taking advantage of a new feature of nature unlike other animals. This new feature is the same though the sources of food or geo-climatic conditions are different. The method is always to take advantage of the repetitive nature of the external events. Hence the selection pressure on the mutant genes is the same, in the direction of increasing memory power and thereby brain matter. The spread of such genes throughout mankind ensured a uniform genetic code to the entire mankind. This is the basic genetic endowment of mankind.

There are however some racial differences like skin colour, colour of eyes, and hair, shortness or tallness, hairiness or lack of hair, kinkiness of hair, shape and size of head, nose, ears, eyes as well as differences in frequency of blood groups in different races. However they do not in any way alter the basic genetic endowment of mankind which gave man his uniqueness. The difference between the genes of man and chimpanzee is 1.5%. Then what is the difference between humans of different races. The genes of different races of man are common to an extent of 99.99 per cent. The 0.01 per cent account for racial differences.

Such differences are skin deep. Thus mankind remained a single species though there are several races. He could live in Arctic ice as well as tropical jungles. What is more the two can happily mate together and produce children. In other forms of life such differences would result in different species and reproduction between them is not possible.

Thus the new method man took advantage of to evolve is the cause and effect relationship present in nature, to lie in anticipation of an event as soon as one observes a cause. When this is extended from immediate situations to the observation of movements of the sun and moon and seasons and the availability or scarcity of food, he has mastered the natural events. It is only one step away from observing the seasonal abundance or scarcity to observing how it could be grown by way of grain or how the animals could be domesticated to give milk and meat. Agriculture and animal husbandry are the next logical steps. They developed in Levant by about 10 to 12 thousand years ago at Jericho on the banks of Jordan river and at Jarmo in Iraq the oldest archeological sites unearthed which show clear evidence of agriculture at this time. Then it would be a matter of time before its spread to other areas and it did to river valleys of Nile, Euphrates, Tigris, Indus, Yangtze and Yellow river etc.

Thus, man established himself as the most dominant animal by taking advantage of a feature of nature *i.e.*, the cause and effect sequencing of nature. The pattern present in nature. Nature is not chaotic after all. This enabled him to impose his will not only on animals and plants in hunting and gathering societies but also enabled to impose his will on nature itself by way of subduing it wherever he could and bringing it to his service. This has set him on the path of civilization.

This uniform genetic code of mankind gave a preprogrammed brain to man which constantly searches for cause and effect relationships in nature. If he does not find such relationships his reasoning breaks down and such a breakdown throws him into neurosis, a disturbed condition of mind. Man needs a cause for every effect to keep him in mental balance.

However the story is not that simple. Nature is more clever for man to be taken advantage of and to be subdued. It throws a spanner into his works at the same time enabling him to take advantage of it. Nature exhibits not only causal sequencing, it also exhibits stochastic randomness. That is, breaks exist in the order of things in the sequencing, particularly in individual micro events, though not in large macro events. The sun always rises in a particular direction and sets in the opposite direction with a period of oscillation. That is a macro event. It is not possible in the short span of human life or even the history of mankind to observe deviation from this. Perhaps if one lives billions of years, one may be able to observe deviation in the case of sun also. Man's life span being short, he should be rest content with observations of events which have shorter life spans, perhaps with the behaviour of animals, birds or crops or rain etc. They do not always behave as he expects them to behave, *i.e.,* they jump sequencing. An animal might escape the trap of a hunter, thus disappointing him. Of course, if every time he sets up a trap and an animal is cought in the trap always without fail, then there would be no anticipation and no evolution of mankind. Sometimes no animal is caught in the trap. This introduces anticipation, anxiety and if this happens too often, it produces a scare in the hunter and he seeks reasons for it. He seeks a cause for this irrational behaviour on the part of the animal. Having found none, he invents one. That is a spirit. Now his world is full of spirits. A spirit for a tree, for a bush, for animals, birds, brooks and rivers, mountains and even other men. The invention of the world of spirits is the beginning of religious stirrings in him. He seeks a reason for every event that he could not understand and find a cause. So he supplies the cause in the form of spirits.

Man was compelled to find spirits which latter developed into gods and much latter into one god because he had to find a cause for every event, particularly individual, microlevel stochastic events. He was compelled because he could feel comfortable and assured only with events for which there is a known cause and where he could anticipate. Whenever there is no such known cause and effect relationship could be readily found, he gets confused, bewildered and at a loss and gets neurosis because cause and effect relationship is the one which helped him to evolve and consequently his brain is attuned and preprogrammed by his genetic make up to understand and feel at home with cause and effect relationships and nothing else. However this world is not only full of cause and effect relationships, it also has statistical and stochastic or random effects, which forces man to discover cause and effect relationships where there were none. For this purpose man invented spirits first to provide causes for individual events, latter gods for groups of similar events and much latter one god to supply the unknown cause of all events and he ultimately became the first cause, the cause of all causes. We find evidence of this spiritual world in the reverential burial practices of Neandertals. We find it even more so in the burials of Cro-Magnons, their venus figurines, their art in inaccessible and dark caves which were meant for ritual. Already man was beginning to worry about the unknown. His consciousness is developing. He is yearning to have a connection between himself and the outside world. He wants that connection to be a friendly one. He is developing the stirrings of religion.

The dependence of man on the causation present in nature gave him reasoning faculties. The randomness which is also present in nature gave man awareness, consciousness and self-consciousness which have set him on the path of searching for truth, beauty and goodness, the basics of religious and spiritual life. Without them man would not be human but a glorified ape. This is the uniqueness of human evolution.

14

CONCLUSION

What are the lessons we could learn from this brief survey of the history of the universe as well as that of life on earth and its evolution and particularly human evolution. Does it offer us any guidance to the problems we are likely to face in the 21st century. Yes, I think it does provide us some guidance to order our affairs during at least the first 30 years of 21st century. So let us briefly recapitulate the gist of the preceding chapters and see how we could learn something from them.

First, man is a part and parcel of this universe. He, as well as all the flora and fauna and all other living things as well, are a part of this universe. He as well as every thing else on this planet earth are made up of atoms of elements cooked up in stars and latter spewed up into the space as the stars disintegrated. Only Hydrogen and Helium are the primary elements formed out of the cooling radiation of the 'big bang'.

We are alone on our Planet Earth, the only intelligent life in our solar system. There may be intelligent life forms in our Milky Way galaxy or in other galaxies. We have found none so far in spite of efforts to make contacts with such intelligent life forms. The hope is also slim. The distances involved are so great and the transmission of radio signals to and from such intelligent life in our own galaxy or other galaxies is full of many difficulties. So we have to live alone on our Planet Earth at least until some breakthrough is achieved.

Man can live only on this Planet Earth since his evolutionary past has made him so. Is there then any plan or purpose behind man's evolution. We can find none from the facts of evolution. Mutations in genes, the chief architects of evolutionary change occur at random and not to order or by previous design. The right mutant gene at the right time is selected by nature itself and propagated through the population. Many organisms led to blind ends. Some organisms preserved some features which allowed for further evolution. This process has finally led to Homo Sapiens some 50 thousand years ago and we are all their descendants.

If man wants to live in the outer space or on moon or any other planet he has to create conditions that simulate conditions on Earth. It is unnatural, highly expensive and possible only for limited periods, for only a few for experiments. We already know that those who are returning from space stations after spending a few months there are suffering from physical and mental disorders. The Soviet space station MIR completed its life as an experimental station and it was brought down to the Earth's atmosphere and burnt down in a controlled manner over the Pacific Ocean. Another International space station is under way with the joint cooperation of U.S.A and Russia. The break up of Columbia space shuttle had cast a shadow of doubtful scrutiny on the necessity of this space station. Large scale emigration of earthlings into outer space or to other Planets is out of the question. That makes no sense. The planet Earth itself is our home and spaceship in this unimaginably vast cosmos. Why should we want to build artificial space stations as future colonies for us. It would be a vain and foolish effort. What we should be doing is to improve the conditions of human estate as well as that of our planet Earth here and now.

Life as we know it exists only on our Planet Earth. Scientists suspect that there might be some kind of primitive bacteria like life on the planet Mars as well as on one of the moons of Jupiter. However all life is kith and kin on our planet Earth. The proteins that make up the various tissues of living organisms are made up of some 20 basic amino acids. The amino acids are ordered by 4 basic nucleotides strung up in pairs of different sequences on the long chromosome chains which are unique for each species and each individual living organism. If life exists on other planets, however primitive that might be, it would be interesting and instructive to know if they also follow the basic pattern of life on Earth. If a different pattern is followed then the origin of life could be somewhere in the cosmos.

The conditions on earth did not remain the same since its birth some 4.5 billion years ago. They changed. Living organisms also changed to suit themselves to the changing outside conditions. There is a mechanism for this change which we now call evolution. The genes *i.e.,* chromosome parts are responsible to produce different proteins *i.e.,* different characteristics in the body. The mutations in the genes alter these bodily characteristics. If these alterations are helpful in the organism's survival, they are retained and passed on to the offspring. If not, they are discarded. Such organisms die out before they have a chance to reproduce. Over long periods, the changes which were selected by nature accumulate in the population and new species are formed to exploit the new ecological niches *i.e.,* environmental conditions.

At each stage of evolution some generalized features were found to be advantageous and retained by the succeeding forms of life. Thus fishes evolved a lower jaw and teeth in it. The amphibians evolved quadrupedal motion and came out of the sea to exploit the land masses. The reptiles overcame the limitations of reproductive restrictions of amphibians which tied them to the water. They evolved the amniotic egg and scaly skins which allowed them to roam the land masses freely without fear of desiccation and they evolved into many species. The mammals evolved temperature control of their bodies which allowed them to be active even in cold climates and at night when there is no sun to warm their bodies. What is more, they gave birth to a few young ones at a time and nursed them with milk and took care of them until they were able to fend for themselves. This reduced the mortality rate among the mammals. Man has retained all these evolutionary innovations and profited from them. On his own part he evolved some novel features which proved to be far reaching and set him apart from other animals not only in degree but also in kind.

Man evolved in the final stages of the evolution in the last 5.5 million years when his stock separated from chimpanzee stock. He evolved into several hominid forms and finally one of them led to Homo Sapiens and the others became extinct. Of this long span of 5.5 million years the early hominids took nearly 3.5 million years to evolve erect posture and free hands with opposable thumb which gave him a power grip as well as a precision grip. He made crude tools chipping away a few flakes from one end of a pebble that fits into the hand and making a sharp edge on the chipped end.

Making of crude tools is the beginning of culture. The tools appeared about 2.0 million years ago and are believed to have been made by a hominid which is now called Homo habilis. His brain capacity is about 770 c.c double that of chimpanzee. During the period between 3.5 million years and 2 million years the physical evolution in Homo is fast. He grew in size and developed larger brains. Large brain gave him conceptual thought which is the basis of culture. Between 2 million and 1.5 million years the evolutionary pressures in Homo habilis stabilized and we have a hominid by 1.5 million years in East Africa which we now call Homo Erectus. He is the harbinger of latter Homo Sapiens. He is not quite human but on the sure path of evolving to full status as a human. From here on his cultural development is faster than his physical evolution. His brain capacity was 900 c.c. and it gradually developed up to 1350 c.c by 100 thousand years ago. He spread out of Africa to Europe and Asia and China.

The evolution in cultural spheres of man during the last 500 thousand years is rapid. His brain capacity has increased to an average of 1350 c.c. His communicating powers through symbolic sounds and ultimately through rudimentary language have developed. This requires conceptual thought. Man took advantage of the causality present in nature to evolve conceptual thought. Larger brain allowed him to have more brain matter and more brain matter allowed him to remember the repetitive causality of nature better and thereby profiting from it, anticipating the sequence of events. It also allowed better coordination in his brain between the associational areas of his various organs particularly his audio-visual and locomotor limbs. Larger brain also permitted cumulation of experiences from one generation to the next. This permits vast development in culture and in due course of civilizations. Man learned to live in larger social groups cooperating with each other as his cultural and civilizational faculties improved.

This dependence on repetitive causality for the final stages of evolution of mankind had given him a uniform genetic endowment because this repetitive causality is of similar nature all over the planet earth giving similar selection pressure to the evolving homos towards having more and more brain matter. There are however some superficial differences in his physiognomy depending upon local atmospheric conditions and geography. These produced different races and different cultural adaptations in mankind, but not different species. These racial differences do not effect the basic human genetic endowment which is the same all over the world, which distinguished man from other animals. By basic human genetic endowment we mean that he has stereoscopic color vision, erect posture, opposable thumb which allowed him to make and operate delicate tools and instruments, developed conceptual thought and communicate through language, his experiences are cumulative, lives in social groups cooperating with each other dividing labour and exchanging the products of such labour.

As a result of his long memory he also developed awareness of his surroundings as well as about himself to a greater extent which was not found in other animals. This had led to consciousness as well as self-consciousness and questions for which there are no ready answers. He had to find the answers himself. There are no answers in nature. Nature is impersonal and unconcerned. This led to a search for truth, beauty and goodness. As a legacy of his past evolution he also has passions some base and some noble. Mankind is tormented by the conflict between base and noble passions. That in short is what I mean by the basic human genetic endowment. This is common for all homo sapiens on this planet Earth irrespective of race, for both males and females.

As a result of his evolutionary past mankind has also accumulated some problems. They are in short the dominance of man among all organisms on this planet Earth and as a consequence he had proliferated in numbers to the detriment of all other organisms as well as plant life. Secondly he has developed into a mosaic of races and cultures over the face of the earth. Though there is a genetic unity there is diversity in their expression in the form of various racial shades and cultural adaptations and innovations. These are producing some tensions. Thirdly the base and noble passions of man which are at perpetual conflict within each man and again within each society and between societies are threatening to disturb peace nay even much more. These three basic problems and their consequences have acquired threatening proportions during the second half of 20th century though they were with us for many centuries. They were under check and control for various reasons. But now they have acquired alarming proportions which man could ill afford to ignore. He has to spend the first half of 21st century in finding solutions for them. The best energies of the generation which is now in its 20s and 30s have to be devoted to find solutions to this legacy of the second half of 20th century. Otherwise there may not be much future for mankind as a viable organism on this planet earth. We will now consider them briefly.

All organisms in nature do not live up to the age of reproductive maturity and reproduce. Many organisms become food for their predators and die. Man alone does not become food for other organisms. He kills himself in wars and genocides. Famines and natural calamities like earthquakes, cyclones and floods take their toll. Above all bacteria and viruses take a deadly toll. The latter two causes which were the major killers of humans were largely controlled during the second half of 20th century by the development of better medical and health care practices. As a result the population of the world increased from 2.5 billion in 1950 to 6 billion by 2000. That is one problem. The uneven distribution of population among the countries of the world is a sub problem. In the already well developed countries of Western Europe, Japan and Russia the population is decreasing. In the under-developed nations the population is increasing. This is resulting in large scale transfer of population from undeveloped countries to Western Europe and the United States of America. This is creating some tensions in those societies. This has to be tackled with imagination and proper planning for the transfer of skilled people from surplus countries to deficit countries.

The large scale transfer of people from India, China and other Asian countries requires these countries to invest in human resource development so that their level of education and other skills are improved to the level where the recipient countries would welcome them. Their health and nutritional standards also have to be improved by widespread immunisation, provision of potable drinking water, provision of decent latrines and other sanitary and personal hygiene facilities. At least 10% of the national GDP has to be invested on the future generation. Otherwise no country can hope to better its lot. It is an infrastructure development expenditure which yields better results than investments in dams, roads, electricity, ports, railways and telecommunications. When the people themselves are not in a position to use these infrastructure facilities what good are they for. So the immediate need is to invest not less than 10% of GDP in subsidised health, nutrition and human resource development departments. Unfortunately that does not seem to be the case.

As a corollary to this population growth, there are rising expectations of the people of under-developed countries for better standards of life. This puts a pressure on the limited resources of the earth. The advanced industrial nations of Europe, America and Japan are using up more natural resources of the planet Earth than the vast population of the rest of the world. If their living standards also were to be increased to the level of the industrialised nations then there simply do not exist the kind of resources necessary on this planet. Either we have to use the resources in a more efficient, better and more equitable way or we have to curtail our appetite for more resources *i.e.*, for better living standards. This could be done only by a more equitable distribution of resources by a reallocation of resources from nations which are wasting them in an extravagant manner to nations that are struggling even to meet basic necessities of life.

There is an increasing imbalance in the living conditions of the people within the nations as well as between the nations.

In the industrialised nations these imbalances between individuals is mitigated to some extent by social security measures. The imbalance in living conditions in the developing nations is of a different type. Here it exists between those who are in the organised sector and those who are in the unorganized sector. The former enjoy all securities and benefits and the latter none at all.

There is a vast difference in the living conditions of the industrialised and the developing nations. The former are the guzzlers of the natural resources of the Earth and also its major polluters. They are trying to enjoy a sort of economic colonialism over the latter which the latter are resisting with varying degrees of success. Consumer economies which are being developed now in the name of Free World Trade will only increase the demand for resources and ultimately lead to more tensions.

Some people however argue that the development of science and technology will permit the extraction of resources now locked up beneath the seas or beneath the ice cap of Antarctica. We could extract energy from Sun in a pollution free way. We could find new uses for old and discarded materials and new technologies for more efficient use of resources etc. These are of course all in the future. More extraction of resources mean more pollution. Ultimately mankind has to learn to live within the means available to the population of the entire globe. It can be done only when each individual and nation sets up limits to their consumption. This can not be done at the expense of others within the same nation or at the expense of other nations. That will only lead to tensions and disturbance of peace. Witness the demonstrations at the meetings of WTO, WORLD BANK and the heads of state of industrialised nations. They will only increase further in the future.

The burning of coal and oil for industrial purposes and the automobile exhausts are releasing carbondioxide into the atmosphere, thus increasing the green house gases. The use of C.F.Cs in refrigerators, air conditioners and the spray cans etc., is releasing them into the atmosphere and they are reacting with the ozone layer and depleting ozone layer thus allowing free passage of ultraviolet rays into the earth's atmosphere.

Thus, the green house gases and the ozone depletion are increasing the earth's temperature. The cutting up of forests is reducing their capacity to absorb the carbondioxide. All these causes are contributing to the warming of earth's atmosphere and causing imbalance in the earth's wheather conditions which is resulting in severe droughts, floods, cyclones and other calamities. The heat wave conditions in Europe in July-August of 2003 resulted in the death of nearly 23 thousand people due to illnesses related to heat. Of these France alone accounted for half the cases.

The use of fertilizers and pesticides, nuclear wastes generated in power stations and the variety of wastes generated in the industries are creating pollution and degradation and imbalance in the ecology and environment. Man himself is a major polluting animal. The removal and disposal of his bodily and domestic waste has become a problem to the civic bodies all over the world.

The increasing concentration of the population in metropolitan areas is creating major problems of how to provide safe drinking water, unpolluted air, acceptable accommodation in living units and the attendant health, medical and educational and mass transportation facilities are increasingly becoming difficult. These problems are creating urban slums and ghettos, crime and violence, addiction to drugs and drink, alienation of family and in short a dehumanizing atmosphere is being created.

So the population problem has to be tackled by a transfer of people from surplus states to deficit states. The deficit states are not whole heartedly welcoming such a move though ultimately it becomes inevitable. The surplus states have to invest heavily in health, nutrition and education of its growing up generation if they have to cash in on this opportunity. If the one child norm of these deficit states catches up with the surplus states also then the new generation will not be sufficient to replace the dying out generation. Population crunch will result. China is officially following one child per couple norm with some relaxation in rural areas. In India the northern states are producing more children and the southern states less. Already some families are following one child norm.

This may not be good in the long run. Two children per couple is a proper norm.

A corollary to this population problem is the resource crunch. If every one expects to reach the living standards of advanced nations then the world as a whole do not have such resources. Ultimately every individual and every nation has to set up limits to their consumption norms. An attendant problem of extracting and consuming more resources means creation of more pollution and degradation of environment and ecology. An equitable and just balance has to be found between population growth, rising expectations for better living standards and depleting resources and environmental degradation and pollution.

The spread of Homo Sapiens to the four corners of the habitable earth exposed them to varying environmental conditions which necessitated adaptations to suit those conditions. This gave rise to various cultural varieties. Further these ecological differences produced several different racial features which are skin deep only. The basic human genetic endowment of mankind irrespective of race is same. This realisation was quite recent. Its erroneous understanding gave rise to much misery in the world in 20th century. The cultural differences are also due to the necessity to adapt to the varying ecological conditions. Apart from the basic need of obtaining food there is a large part of culture in the form of ritual, ceremonies, mythologies behaviour codes of morals and ethics, objects of worship and prayer. These resulted as man developed awareness of the world outside and its behaviour, self-consciousness and his own needs and wants and to answer the larger questions of wherefrom we have come, what is the purpose of life, is there a purpose behind this, if so what is the relationship between the self and non-self. What is the nature of non-self and its behaviour. Man wants to be on friendly terms with this non-self. So propitiations and prayers, a god and religion developed. These have taken various forms in various cultures but the biological need and the genetic background for their development is the same.

Molecular biologists propounded two lines of thought which point out that modern man, Homo sapiens or Cro-

Magnons originated somewhere in Africa around 60 thousand years ago and then spread throughout the world. They point out the mitochondrial Eve and Y Chromosome Adam as support to their arguments. Mitochondria is a cell body found outside the cell nucleus and between the cell wall. It has its own D.N.A. It passes on from mother to daughter. The Y Chromosome is responsible to give male sex to the child. It passes on from father to son. By the amount of mutations present in these D.N.A. molecular biologists built up a time chart. By this chart they estimated that Cro-Magnon man had originated in Africa some 60 thousand years ago and spread throughout the world by 30 thousand years ago.

On the other hand, anthropologists in particular, point out that Homo Erectus had spread out of Africa as far back as a million years ago and they have evolved into local races. He never isolated himself but passed on useful genes from one geographical locality to another thus maintaining genetic unity in spite of racial and cultural diversity. Some groups probably the Neandertals might have evolved in isolation due to ice ages and finally became extinct some 40 thousand years ago. Whether they contributed any of their genes to the latter Cro-Magnons or not is an undecided point. Thus molecular biologists and anthpologists present two pictures for the last stages of human evolution. There is no agreement yet.

Thus the world is now a mosaic of races and cultures. Yet there is a basic unity in the genetic endowment of different races. Geographically proximate races merge into one another imperceptibly. Only when they are separated by wide geographical dividers like mountain ranges and vast oceans they show perceptible differences. Similarly the cultures also do not exhibit perceptible differences in geographical proximity. Only when they are divided by geographical barriers they show cultural differences. These again show some basic answers to human necessities and the preprogrammed pattern imposed on humans as a result of their uniform genetic endowment. Thus the racial differences are only skin deep and do not effect the basic genetic endowment of humans. Similarly the cultural differences also are superficial basically arising out of their necessity to obtain a living by adapting themselves to their environment and be on friendly terms with that environment and impose an order in their society. These then are similar needs though they produced different cultures. So the cultural differences are also superficial.

Man is also heir to many passions as a result of his evolutionary legacy. Some of these passions are base and some noble. The base passions express themselves as lust, anger and wrath, covetousness and greed, pride and prejudice, jealousy and vengefulness, etc. The noble passions often are expressed as love, charity, fellowship, empathy, cooperation, pity and compassion and peace, an intense desire for order in society by moral and ethical behaviour, a search and longing for truth, beauty and goodness. There is always a struggle within each man as well as within each society between these opposing passions. The base passions will take the upper hand in some people and the noble passions take the upper hand in some people. Circumstances, times and events also play their role in deciding which passions gain the upper hand. In this struggle some people suffer loss of property, loss of freedom or even loss of life. Such sacrifices often bring out more awareness to the other people about the noble passions for which the persons sacrificed and their sacrifice and loss will become gain to the succeeding generations by influencing their behaviour in favour of the noble passions. Jesus Christ was crucified. Abraham Lincoln was shot dead and so were Mahatma Gandhi and Martin Luther King. But their death was not in vain. They changed the lives of many of their succeeding generations towards following the noble passions. Both the base and noble passions exist in mankind as potentialities. The events and prominent people of their times bring these potentialities to reality and expression.

Unfortunately the 20th century was dominated by base passions of mankind as a legacy of the events of 19th century colonialism and the early evils of industrialization. Man had learned how to subdue nature to yield her bounty as well as how to harness her power. This control over nature led to exploitation and sometimes over exploitation of nature's resources, the minerals, land, air and water. Subjugation and exploitation of groups of people within the same nation as well as other nations and colonies followed. This led to uprisings and upheavals within the nations as well as war between nations. The 20th century saw two great wars apart from many smaller wars and many genocides and a long spell of 45 years of cold war. All these have effected the lives of people profoundly and we at the beginning of this 21st century are faced with some problems as a legacy of 20th century.

Planet Earth is now threatened by the pile up of nuclear weapons by the big five nations and some other countries are also surreptitiously or otherwise acquiring nuclear weapons capability. India and Pakistan have already acquired this capability. Added to this is the danger of developing biological and chemical weapons.

Then is there any safety for mankind, nay, for life on earth with all this destructive power scattered all over the world. We have to consider the moral questions involved in the development and continued deployment of such dangerously destructive power and its continuation any further. There are no valid reasons other than the big power ambitions to have a big brotherly say and hegemony in the affairs of the world.

The nation states and the societies they are supposed to guard are faced with terrorists in the state or from outside the state. These groups are trying to create acts of violence at a particular place and time often vulnerable, at their choice and thus browbeat the governments to grant their demands. If the same has some support from the people, then the government and its enforcement agencies are faced with an uphill task. On the other hand if the terrorists do not enjoy support from the people, they in due course are trying to terrorise the people also to get what they want. These groups often have their beginning when some of them feel they have no say in the decision making process of the state.

Often the terrorists as well as the repressive arm of the governments commit acts of human rights violations. There is no use accusing one another of these violations. It often boils down to the pot calling the kettle black. Both sides are equally blame worthy. There is no gain accusing any party as guilty of violating human rights. One should rather work for ending the conditions which are the breeding grounds of insurgents and terrorists.

Modern world is facing the problem of drug addiction and horrible diseases like AIDS. The alienation of man from society is perhaps driving him to acts of violence as well as addiction. These are perhaps more psychological, social, economic and political problems than mere problems of addiction. There is an inalienable connection between drug traffickers and terrorists and criminals. The growers of the base drugs opium and charas are the poor farmers of Asia and Latin America. Then they pass on into the hands of international organised criminal groups which are often having close connections with leaders in other walks of society, often politicians and the enforcement agencies. This is making the problem all the more sinister. Purified and refined drugs are often being used as international hot currency by terrorist groups and unscrupulous government agencies that have connection with them. This is posing additional problems. Thus drugs, terrorism and human rights are all different facets of the same problem. That could be controlled only if we identify the problem and address ourselves to solving it. But the present efforts of the governments seem to be to turn a blind eye towards the problem and try only to suppress the effects using the state machinery. This is not solving the problem but merely driving it underground.

Modern medicine, in spite of many admirable advances is also working to reduce the vigour of human genetic endowment. It is permitting many individuals to live up to adulthood and then beget children, though some of them would have died in the infancy itself due to several diseases that could be due to genetic defects. This is permitting those parents with defective genetics to pass them on to their offspring. This process is resulting in the depletion of the vigour in the gene pool which ought to be there if natural selection was permitted to weed out the weaklings. This will effect the vigour of the human gene pool in the long run. So it is desirable to see that those with genetic defects that could be passed on to their children do not procreate. We should use such controls with full knowledge and responsibility.

The advocates of eugenics a century ago in the inadequacy of their knowledge attracted much odium towards such methods. Such mistakes should not be repeated.

Lastly there is the problem of information control, distortion and doctoring and false propaganda by interested groups and governments to gain control over the public opinion. In democratic countries this is being done by interested groups as well as government agencies in a covert manner. However they can not get away with their mischief forever. In totalitarian countries this goes on in a brazen fashion by means of censor and state propaganda. Man, after all forms his opinions and behaviour patterns based on the information at his disposal. If false information is being pandered overtly or covertly, then he is helpless and in due course succumbs to such false propaganda.

Even now some hate groups are conducting their propaganda all over the world and in all the countries irrespective of those countries being democracies or dictatorships. This is leading to the vicious circle of false propaganda, organized terrorism, human rights abuse, drug trafficking, gun running, money laundering and such other abuses which are posing a danger to the modern world.

These then are broadly the problems faced by each and every society to some degree or the other. Some societies may feel the effects of some problem more acutely than the others but no society is free from these problems. Without cooperation and concerted action between these several societies which are now more or less organized into nation states, none of these problems could be solved. Now the society has become global though there are many nation states in it. Travel by supersonic and Jumbo Jets and communication through satellites had made the concept of global village a reality. We are forced to think globally and we are also forced to think how each society is interlinked and interdependent on each other. No society could live any longer in isolated seclusion. The problems faced by the various nations of this world could be solved only if we evolve a global society that cherishes some common values based on knowledge, equity, justice and compassion followed by action which is concerted and responsible. However the nations of the world are not yet working in this direction. They are still in the narrow mind set of the 20th century.

The future of mankind as well as the future of this planet Earth will depend to a large extent on how best man could use his unique capacities to promote his collective well being by a balanced and judicious use of natural resources of this planet Earth. That requires development of proper social, moral, legal and government forms for it and the necessary ideological framework to sustain it. That is the predicament of our age, the age that had spent its adult life in the post Second World War years and the age that is going to step into the first one or two decades of 21st century as elderly persons and hand over the baton to our successors. They have to formulate a world view based on the accretion of knowledge during 20th century, when the base passions of mankind dominated the course of history. In the 21st century the noble passions of man have to take the upper hand.

Two generations in the first half of 20th century were denied this opportunity by two senseless wars. It has at last fallen to the last generation of the century to sum up the developments of the century and extract from it the knowledge and wisdom we have to hand over to our successors. Handing over a better world to our successors than what we have inherited from our predecessors is the foremost virtue and ethics. This is the lesson we have learned from such a summing up of history, particularly the history of the 20th century.

BIBLIOGRAPHY

1. General

Attenborough, David. *Life on Earth: A Natural History.* The Reader's Digest Association Ltd., and William Collins and BBC, 1980.

Bronowski, J. *The Ascent of Man.* Futura Publications, 1973.

Colbert, Edwin H. *Men and Dinosaurs.* Penguin Books, 1968.

Du Nouy LeComte. *Human Destiny.* Mentor Books, 1947.

Flanagan, Dennis. *Flanagan's Version.* Affiliated East-West Press Private Limited, 1989.

Frisch, Karl von., Lowenstein, Elsa B. (Tr) *About Biology.* Oaliver & Boyd.

Gardner, Martin. *Great Essays in Science.* Washington Square Press, New York, 1957.

Greene, John C. *Darwin and the Modern World View.* Mentor Books, 1963.

Harre, R. *The Philosophies of Science.* Oxford University Press, 1972.

Huxley, Julian. *Man in Modern World.* Mentor Books.

John Palmer (Ed). *Exploring the Secrets of Nature.* The Reader's Digest Association, Inc of Pleasantsville, New York, U.S.A.

Moore, Ruth and the Editors of Time-Life Books. *Evolution,* 1962.

Moorehead, Alan. *Darwin and the Beagle.* Penguin Books, 1971.

Newman, James R. *What is Science.* Washington Square Press, New York, 1961.

Simpson, George Gaylord. *The Meaning of Evolution.* Oxford & I.B.H. Publishing Co. 1965.

2. Cosmos

A.D'Abro. *The Evolution of Scientific Thought from Newton to Einstein.* Dover Publications Inc.

Bergamini, David and the Editors of Time-Life Books. *The Universe,* 1962.

Cowen, Pon., *Galaxy Hunters.* National Geographic, Feb. 2002.

Feldman, Anthony. *Secrets of Space.* Aldus Books Ltd., London, 1980.

Hawking, Stephen W. *A Brief History of Time.* Bantam Books, 1988.

Huens, Jean Leon, Corly, Thomas Y. *Seven Men Who Solved Riddles of the Cosmos.* National Geographic, May, 1974.

Lovell, Bernard and Joyce. *Discovering The Universe,* The English Language Book Society and Ernest Benn, 1964.

New Cott, William R. *The Hubble Telescope Views the Universe from Space.* National Geographic, April, 1997.

Owen, David. *Into Outer Space.* New Burlington Books, 2000.

Sagan, Carl. *Cosmos.* Futura Macdonald & Co., London, 1981.

Weaver, Kenneth F. *The Incedible Universe.* National Geographic, May, 1974.

3. The Earth

Beiser Arthur and the Editors of Time-Life Books. *The Earth,* 1970.

Engel, Leonard and the Editors of Life. *The Sea,* 1961.

Miller, Jr. G. Tyler. *Living in the Environment.* Wadsworth Publishing Co., Belmont, California, 1994.

4. Origin of Life

Burnet, Macfarlane. *Genes, Dreams and Realities.* Penguin Books, 1973.

Crick, Francis. *Life Itself.* Futura Macdonald & Co., 1982

Hoyle, Fred and Wickramasinghe, Chandra. *Evolution from Space.* Granada Publishing Co., 1983.

Moore, Ruth. *The Coil of Life.* Dell Publishing Co., 1983.

Watson, James D. *The Double Helix.* Penguin Books, 1970.

5. The Process of Evolution

Altenburg, Edgar. *Genetics.* Oxford & I B H Publishing Co., 1970.

Dobzhansky, Theodosius. *Evolution, Genetics and Man.* Wiley Eastern Private Ltd., 1971.

Dobzhansky, Theodosius. *Genetics and the Origin of Species.* Oxford & IBH Publishing Co., 1951.

Dodson, Edward O. *Evolution : Process and Product.* Affiliated East-West Press Pvt. Ltd., 1968.

Goodal, Jane Van Lawick. *In the Shadow of Man.* Fontana/ Collins, 1973.

Handler, Philip. *Biology and the Future of Man.* Oxford University Press, New York, 1970.

Hooton, Earnest A. *Up from the Ape.* Motilal Banarsidas, 1965.

Huxley, Julian (Ed). *Evolution as A Process.* Collier Books, 1954.

Madawar P.B. and J.S. *The Life Science.* Granada Publishing, 1978.

McKusick, Victor A. *Human Genetics.* Prentice Hall of India Private Limited, 1972.

Muller, Herman J. *Studies in Genetics.* Oxford & IBH Publishing Co., 1969.

Sinnot, Dunn, Dobzhansky. *Principles of Genetics.* Tata McGraw-Hill Publishing Co. Ltd., 1973.

Winchester A.M. *Genetics.* Oxford & IBH Publishing Co.

6. Vertebrate Evolution

Carrington, Richard and the Editors of Time-Life Books. *Mammals,* 1963.

Colbert, Edwin H. *Evolution of the Vertebrates.* Wiley Eastern Limited, 1970.

Hans Querner etc. *New Perspectives in Evolution.* Transworld Publishing Co., 1973.

The Editors of Reader's Digest. *Curious Creatures,* 1981.

The Editors of Reader's Digest. *Secrets of the Past,* 1980.

7. Primate Evolution

Eimerl, Sarel and De Vore, Irven and the Editors of Time-Life Books. *The Primates,* 1965.

8, 9, 10 and 11 Human Evolution

Berger, Lee. *Redrawing Our Family Tree.* National Geographic, August, 1998.

Buettner-Janusch, John. *Origins of Man.* Wiley Eastern Private Ltd., 1969.

Burkitt, M.C. *The Old Stone Age.* Rupa & Co., 1977.

Clark, Grahame. *World Prehistory.* Cambridge University Press, 1977.

Day, Michael H. *Fossil Man.* Hamlyn Publishing Co., 1969.

Gore, Rick. *Expanding Worlds.* National Geographic, May 1997.

Gore, Rick. *Neandertals.* National Geographic, January, 1996.

Gore, Rick. *New Find.* National Geographic, August, 2002.

Gore, Rick. *People Like Us.* National Geographic, July, 2000.

Gore, Rick. *The First Europeans.* National Geographic, July 1997.

Gore, Rick. *Tracking the First of Our Kind.* National Geographic, September, 1997.

Haviland, William A. *Human Evolution and Prehistory.* Holt, Rinehart and Winston, 1974.

Howell F. Clark and The Editors of Time-Life Books. *Early Man,* 1970.

Johanson, Donald C. *Face to Face with Lucy's Family.* National Geographic, March, 1996.

Leakey, L.S.B and Goodall, Vanne Morris. *Univeiling Man's Origins.* Methuen & Co Ltd., 1970.

Leakey, Meave. *The Farthest Horizon.* National Geographic, 1995.

Leakey, Richard and Levin, Roger. *Origins Reconsidered.* Little Brown and Company.

Leakey, Richard and Walker, Alan. *Homo Erectus Unearthed.* National Geographic, November, 1985

Marshak, Alexander. *An Ice Age Ancestor.* National Geographic, October, 1988.

Marshak, Alexander. *Exploring the Mind of Ice Age Man.* National Geographic, January, 1975.

Putman, John J. *The Search for Modern Humans.*

Rigaud, Jean, Philippe. *Art Treasures from the Ice Age Lasca'ux Cave.* National Geographic, October, 1988.

Weaver, Kenneth F. *The Search for Our Ancestors.* National Geographic, November, 1985.

Young, J.Z. *An Introduction to the Study of Man.* Oxford University Press, 1974.

12. The Role of Brain

Bailey, Ronald H and the Editors of Time-Life Books. *The Role of the Brain,* 1976.

Jean Pierre Changeux and Jean Chavaillon (Ed). *Origins of the Human Brain.* Clarendon Press, Oxford, 1996.

Reader's Digest. *ABC's of the Human Body.*

INDEX